Border Regions Series
Series Editor: Doris Wastl-Walter
University of Bern, Switzerland

In recent years, borders have taken on an immense significance. Throughout the world they have shifted, been constructed and dismantled, and become physical barriers between socio-political ideologies. They may separate societies with very different cultures, histories, national identities or economic power, or divide people of the same ethnic or cultural identity.

As manifestations of some of the world's key political, economic, societal and cultural issues, borders and border regions have received much academic attention over the past decade. This valuable series publishes high quality research monographs and edited comparative volumes that deal with all aspects of border regions, both empirically and theoretically. It will appeal to scholars interested in border regions and geopolitical issues across the whole range of social sciences.

Intra-Africa Migrations
Reimaging Borders and Migration Management
Edited by Inocent Moyo, Jussi P. Laine and Christopher Changwe Nshimbi

Invisible Borders in a Bordered World
Geographies of Power, Mobility, and Belonging
Edited by Alexander C. Diener and Joshua Hagen

Borderlands Resilience
Transitions, Adaptation and Resistance at Borders
Edited by Dorte Jagetic Andersen and Eeva-Kaisa Prokkola

Invisible Borders in a Bordered World
Power, Mobility, and Belonging
Edited by Alexander C. Diener and Joshua Hagen

Frontex and the Rising of a New Border Control Culture in Europe
Antonia-Maria Sarantaki

For more information about this series, please visit: www.routledge.com/Border-Regions-Series/book-series/ASHSER-1224

Frontex and the Rising of a New Border Control Culture in Europe

This book examines the rapidly expanding EU agency's distinct role in EU border control, showing that Frontex is a prominent border control actor that reshapes the EU borders by promoting a new border control culture.

Bringing culture into the analysis of Frontex, this book offers an alternative in-depth understanding of the agency's function, focusing on the production and diffusion of border control assumptions and practices within a border control community. Based on data drawn from primary research at Frontex and two EU external borders, namely Lampedusa and Evros, this book examines Frontex's contribution to the emergence of a new border control culture in Europe, replacing the pre-existing Schengen culture. Compared with the existing literature on Frontex, this novel account takes into consideration the evolving nature of borders and border control, discussing three contemporary challenges for the established border control regime: Brexit, the COVID-19 pandemic, and hard security preoccupations, such as the fall-out from the Russian invasion in Ukraine and the weaponisation of migration at the Greek-Turkish land border.

Frontex and the Rising of a New Border Control Culture in Europe will appeal to scholars and students of border management, EU studies, migration, geography, international relations, and security, along with policymakers and practitioners with an interest in EU border control and Frontex.

Antonia-Maria Sarantaki is postdoctoral researcher in the School of Economics and Political Sciences at National and Kapodistrian University of Athens and academic fellow in the Department of International and European Economic Studies at Athens University of Economics and Business. She holds an MSc Econ in Security Studies from Aberystwyth University, UK, and a PhD in European Politics from Panteion University of Social and Political Sciences, Greece. Her PhD thesis focused on borders and the EU border control policy regime.

Frontex and the Rising of a New Border Control Culture in Europe

Antonia-Maria Sarantaki

LONDON AND NEW YORK

First published 2023
by Routledge
4 Park Square, Milton Park, Abingdon, Oxon OX14 4RN

and by Routledge
605 Third Avenue, New York, NY 10158

*Routledge is an imprint of the Taylor & Francis Group,
an informa business*

© 2023 Antonia-Maria Sarantaki

The right of Antonia-Maria Sarantaki to be identified as author of this work has been asserted in accordance with sections 77 and 78 of the Copyright, Designs and Patents Act 1988.

All rights reserved. No part of this book may be reprinted or reproduced or utilised in any form or by any electronic, mechanical, or other means, now known or hereafter invented, including photocopying and recording, or in any information storage or retrieval system, without permission in writing from the publishers.

Trademark notice: Product or corporate names may be trademarks or registered trademarks, and are used only for identification and explanation without intent to infringe.

British Library Cataloguing-in-Publication Data
A catalogue record for this book is available from the British Library

ISBN: 978-1-032-13585-4 (hbk)
ISBN: 978-1-032-13641-7 (pbk)
ISBN: 978-1-003-23025-0 (ebk)

DOI: 10.4324/9781003230250

Typeset in Times New Roman
by KnowledgeWorks Global Ltd.

To the memory of my father, Ilias A. Sarantakis,
and brother, Konstantinos Deliaslanidis.

Contents

List of illustrations x
Preface xi
Acknowledgements xiii
List of abbreviations xv

1 Introduction: The irrelevancy of Schengenland and Frontex's rise   1

Border evolution 3
New 'kids' on the border control block 4
Setting Frontex's scene 5
Border control actors 6
Speaking of culture 8
Cultures of border control 10
To Schengen or not to Schengen? 12
A new border control culture? 14
Book overview 15
Notes 17
References 17

2 Frontex: An insurgent border control actor   23

Introduction 23
Frontex's birth 24
Frontex's governance 25
Frontex's expansion 28
Frontex's institutional brigade 32
Frontex's deeds 34
Frontex's role in EU border control: reviewing the literature 36
Towards Frontex's reinvigoration: bringing culture in 38
Conclusion 40

viii Contents

    Notes 40
    References 40

3 Constructing Frontex's culture   44

    Introduction 44
    Frontex: the essential EU border control actor 45
    Delving into Frontex's cultural traits 50
    Conclusion 60
    Notes 60
    References 60

4 Constructing the border of Lampedusa   67

    Introduction 67
    The island setting 68
    The making of a border 70
    Border control conduct and actors 75
    Frontex on the field 76
    Border control assumptions and practices in Lampedusa 78
    Conclusion 84
    Notes 84
    References 85

5 Constructing the border of Evros   89

    Introduction 89
    The river setting 90
    The making of a border 93
    Border control conduct and actors 97
    Frontex on the field 98
    Border control assumptions and practices in Evros 102
    Evros and Lampedusa: drawing differences and
      similarities 107
    Conclusion 109
    Notes 109
    References 110

6 Border control in process: The rise of Warsaw culture   114

    Introduction 114
    Culture loading 115
    And its name shall be Warsaw culture 119

Schengen, Westphalia, Brussels, and Warsaw: Variant in name only? 122
From Schengen to Warsaw 126
Frontex in Warsaw 129
Out of Frontex's box: EU institutions in Warsaw 138
Conclusion 140
Notes 141
References 141

7 **Challenges to Warsaw culture** 147

Introduction 147
Unsettling a settled culture 148
Brexit and the English Channel 149
Borders amid the COVID-19 pandemic 155
Borders and hard security 161
Conclusion 167
Notes 169
References 170

8 **Conclusion: Frontex's leadership and the re-drawing of EU border control** 175

Evolving borders 176
The rise of a new border control culture 177
The Frontex effect 178
Frontex's cultural impact 181
Between a rock and a hard place? 183
Looking ahead 185
Concluding reflections 186
Note 187
References 187

Index 189

List of illustrations

Charts

2.1	Frontex's staff expansion	28
2.2	Frontex's budget expansion	29
4.1	Irregular sea arrivals to Italy	73
5.1	Irregular mobility via the Greek-Turkish border	94

Figures

1.1	The cultural evolution process	11
2.1	Frontex's organigram in 2008	29
2.2	Frontex's organigram in 2022	30
2.3	Frontex's main tasks	35

Maps

4.1	Map of Lampedusa border	69
5.1	Map of Evros border	92

Tables

3.1	Frontex's border control cultural traits	59
4.1	Irregular sea migration to Lampedusa	72
4.2	Frontex in Lampedusa	78
4.3	Border control cultural traits in Lampedusa	83
5.1	Irregular migration to Evros	96
5.2	Frontex in Evros	101
5.3	Border control cultural traits in Evros	106
6.1	Warsaw border control culture	119
6.2	Cultures of border control in Europe	123

Box

B2.1	Frontex's tasks as listed in its founding mandate	31

Preface

In May 2022, during her visit at Frontex's headquarters, Ylva Johansson, the Commissioner for Home Affairs, described Frontex as the biggest and most important agency that the EU has ever had. On top, Frontex constitutes one of the most well-known EU agencies, as the bulk of EU citizens probably have heard about Frontex and may have participated in strong-opinionated discussions about it. Actually, even third country nationals, whether irregular border crossers during Frontex joint missions at the EU external borders or visa-exempt visitors needing ETIAS authorisation to travel to a Schengen country, know about Frontex, as many of them have been or will be impacted by its activities. Its public recognition – or maybe notoriety – and link to border control render Frontex one of the most highly criticised EU agencies. This can be easily proved by the number of staged demonstrations against it, not only at its headquarters' seat in Warsaw or in other EU capitals, like Brussels, Vienna, and Madrid, but also outside of the EU area, such as in Bern (Switzerland), Dakar (Senegal), and Rabat (Morocco). Regardless of the protests, Frontex continues 'ruling over' the EU border control through continuous mandate enhancements and never-decreasing resources, such as more staff, increased budget, new operational tools, and collective acceptance by its partners. All these enable it to act as a 'primus inter pares' in EU border control, solidifying its privileged position at the EU external borders and in turn stripping away as illusory all calls for its dissolution.

This book, without adopting a pro- or contra-Frontex stance, invites the reader to zoom into Frontex's role in EU border control. In particular, it explores Frontex through a variant prism, namely as an actor shaping both the EU borders and the border control conduct via culture.

To research Frontex and the border control conduct, I visited both the borders and Frontex conducting interviews and fieldwork. That decision proved to be rather challenging involving travelling to remote places, such as in Evros and Lampedusa, with limited resources, and encountering various attempts of blockage from institutional gatekeepers. Overcoming all these challenges was not an easy task but it was a rather pivotal one, as it allowed me to enrich the book's research with valuable primary empirical data. On that account, I hope that the reader will gain a better understanding of the nature of border control conduct, given the scarcity of primary research on border control and especially on Frontex.

My initial engagement with Frontex began 10 years ago, in January 2012, when I commenced a 6-month internship at the Frontex Situation Centre, at Frontex's headquarters, in Warsaw, Poland. Since my early days there, it was not hard to grasp that Frontex was different from the other EU agencies or EU institutions. It was obvious to me that there was a dynamic of rapid growth fostering Frontex institutionally and operationally. That describes a hunch derived by having the opportunity to glimpse inside this EU agency. Actually, I witnessed myself its substantial contribution to the development of the EU borders. The agency was continuously developing applications that were bringing a new modus operandi for border control. It was attracting visitors from border authorities or ministries of third countries, gaining popularity and even admiration for its operational results. Its staff were daily communicating with national border guards, and though in most cases they were not speaking the same language, still they were sharing a common understanding. Certainly, it had also many critics but that was adding to its controversy, at least in my eyes, raising in turn the urge to continue my engagement with Frontex, albeit differently, that is by becoming a research focus in my doctoral and post-doctoral studies.

Besides the insights that I gained about Frontex's role in border control from my internship there, my work at the Greek Asylum Service (from 2016 to 2017, for one and a half years) allowed me to acquire a better understanding of the perspective of border crossers seeking asylum. Actually, my work at the Greek Asylum Service enabled me to obtain a wider picture about migration and reflect upon the impact of border control systems on peoples' lives through coming in direct contact with asylum applicants and witnessing the implementation of national policies on this field, especially in the wake of the 2015 migration crisis.

Hence, my research focus on Frontex is far from ephemeral. Rather, it has had a long gestation, which progressively led to the writing of this book. Indeed, the idea for *Frontex and the rising of a new border control culture in Europe* draws on all these personal experiences, and especially on my doctoral research. Many things have changed since then proven by the emergence of new border control challenges erupting at the EU borders sparking a fresh discussion about the relevance of the pursued border control model. By contrast, others remained the same, such as Frontex's key position in EU border control, as well as the intense migratory pressures at certain frontline countries, like Greece, Italy, and Spain, which amount to a constant both EU and national preoccupation with irregular migration. So, I invite you, the reader, to start the journey of Frontex's role in EU border control by turning the next pages and if you wish to reflect on broader issues of border control, border construction, and migration policies, which never cease to occupy both our minds and our everyday lives, defining cross-border mobility and trans-border transactions.

Acknowledgements

Writing this part of the book indicates that – finally – the time has come for the *Frontex and the rising of a new border control culture in Europe* to be completed. Although book writing is irrefutably a rather lengthy process, a lonely one in case of a monograph, and too stressful (with the deadline fast approaching), I had valuable companions in this endeavour.

    I would like to particularly thank Doris Wastl-Walter, who embraced my book idea from the very beginning, encouraged me to write this book, and accepted it as part of the Border Regions Series. For all these, I shall forever be indebted to her. Next, I would like to warmly thank an anonymous reviewer for the constructive feedback on my book proposal. Special thanks are also due to Dimitris N. Chryssochoou, who I had the privilege to be my PhD supervisor, and since then he has always been encouraging me providing his unfailing support. This book would not have been written (at least in this form) were it not for Ruben Zaiotti developing the 'cultures of border control' analytical framework that inspired my research. A heartfelt thanks goes to all the interviewees for providing their valuable time in sharing their experiences and thoughts with me as well as all the persons that helped me during my fieldwork in Evros and Lampedusa. In addition, I would like to acknowledge the National and Kapodistrian University of Athens that hosted me as a postdoctoral researcher during the time of the book writing. Furthermore, I would like to thank my students at the Athens University of Economics and Business for our stimulating discussions that provided much needed breaks from being in seclusion during the book writing period. I also wish to express my greatest appreciation to Routledge and its editorial team, particularly Faye Leerink, the Commissioning Editor, who showed her support to my project from the first day providing much useful advice and always promptly responding to my emails, as well as her editorial assistants, Prachi Priyanka and Catherine Jones for their invaluable guidance and assistance in the production of this book.

    On a personal note, I am more than grateful to my mother, Eleni Tsouvala, for her patience, understanding, and encouragement, as well as to Dimitris Argyroulis for reading patiently through the manuscript providing critical pointers, bearing my never-ending questions and book anguish, and most importantly for his love and faith in me whenever I had none in myself. Finally, allow me to thank you, the reader, for reading this book and perhaps using it as a source of reflection about the impact of border control. This book is dedicated to all of you and to the memory of my father and brother, who I will always sorely miss.

The completion of this book represents a tangible proof that, regardless of any difficulties or bitter times, we should never abandon our dreams and goals. For me that was the writing and publication of this book, which has now come to a successful fruition.

Antonia-Maria Sarantaki,
July, 2022

List of abbreviations

ABC	Automated Border Control
ADM	*Agenzia delle Dogane e dei Monopoli* (Customs and Monopolies Agency)
AFIC	Africa-Frontex Intelligence Community
BCP	Border Crossing Point(s)
CCTV	Closed-Circuit Television
CEPOL	European Union Agency for Law Enforcement Training
CIRAM	Common Integrated Risk Analysis Model
CJEU	European Court of Justice
CoE	Council of Europe
COVID-19	Coronavirus disease
CSDP	Common Security and Defence Policy
DG	Directorate-General
EBCG	European Border and Coast Guard
ECRE	European Council on Refugees and Exiles
ECtHR	European Court of Human Rights
EEAS	European External Action Service
EFCA	European Fisheries Control Agency
EIBM	European Integrated Border Management
EIBM Strategy	Technical and Operational Strategy for the European Integrated Border Management
EMSA	European Maritime Safety Agency
EPN	European Patrols Network
ESTA	(United States) Electronic System for Travel Authorisation
ETIAS	European Travel Information and Authorisation System
EU	European Union
EUAA	European Union Agency for Asylum (previously EASO: European Asylum Support Office)
EUBAM	European Union Border Assistance Mission
EUROSUR	European Border Surveillance System
EURTF	European Union Regional Task Force
FADO	False and Authentic Documents Online

FLO	Frontex Liaison Office/Officer
FOSS	Frontex One-Stop-Shop
FRA	European Agency for Fundamental Rights
FRAN	Frontex Risk Analysis Network
HM	(United Kingdom) Her Majesty's
IOM	International Organisation for Migration
IT	Information Technology
JORA	Joint Operations Reporting Application
km/km^2	kilometres/square kilometres
LCC	Local Coordination Centre(s)
LIBE	Committee on Civil Liberties, Justice, and Home Affairs
LRIT	Long-Range Identification and Tracking
MAT	*Monades Apokatastasis Taksis* (Units for the Reinstatement of Order)
MS	Member-States
NATO	North Atlantic Treaty Organisation
NCA	(United Kingdom) National Crime Agency
NCC	National Coordination Centre(s)
NGO	Non-Governmental Organisation(s)
OHCHR	Office of the High Commissioner for Human Rights
OLAF	European Anti-Fraud Office
OSCE	Organisation for Security and Co-Operation in Europe
PM	Prime Minister
RABIT/RBI	Rapid Border Intervention Teams/Rapid Border Intervention
SAR	Search and Rescue
SARS	Severe Acute Respiratory Syndrome
SCIFA	Strategic Committee on Immigration, Frontiers, and Asylum
TFEU	Treaty on the Functioning of the European Union
TU-RAN	Turkey-Frontex Risk Analysis Network
UK	United Kingdom
UN	United Nations
UNCLOS	United Nations Convention on the Law of the Sea
UNHCR	United Nations High Commissioner for Refugees
UNODC	United Nations Office on Drugs and Crime
USA/US	United States of America/United States
VMS	Vessel Monitoring System
WB-RAN	Western Balkans Risk Analysis Network
WHO	World Health Organisation

1 Introduction

The irrelevancy of Schengenland and Frontex's rise

> 'Tolerance cannot come at the price of our security. We will defend our borders with the new European Border and Coast Guard'.
> —Jean-Claude Juncker, former European Commission President
> (European Commission, 2016a)

Borders, border control, and migration have become trending terms not only in scholarship but also in political arenas and media discussions. After all, despite the years that have passed since the 2015 migration crisis, border control continues to preoccupy European Union (EU) citizens, given that migration is unstoppably ranked as the most important challenge facing the EU (Eurobarometer, 2019). Or, at least, it was until the outbreak of the COVID-19 pandemic and the war in Ukraine that left Europe's perpetual peace in shatters. Even so, rather than being side-lined in the public debate by emerging security perils, borders gain increasing salience and maintain their relevance. Indeed, the uncertain security landscape in Europe invokes a resurgence of border anxieties about irregular entries and severe migratory flows, impelling us therefore to 'protect our borders' (European Commission, 2021).

Accordingly, in 2019, more than 140,000 persons entered the EU irregularly (Frontex, 2020), whereas 2020 started with thousands of migrants trying to cross the Greek-Turkish land border in Evros after the decision of Turkey to open its doors, triggering fears for a new migration crisis in the EU territory. Such fears were revitalised in August 2021 when Afghanistan was brought once again under the Taliban rule, driving thousands of Afghans to try desperately to flee the country. The same year, the EU started to face increased migration pressures at its eastern border after the sanctions imposed against Belarus. Even at the time of writing (end of 2021 — first half of 2022), the EU is witnessing the largest wave of migration influx in Europe since World War II, spurred by the 2022 war in Ukraine and its decision to adopt an open border policy for Ukrainians. Simultaneously, new border crossing restrictions are being put into place in response to COVID-19,

DOI: 10.4324/9781003230250-1

severely curtailing the fundamental right of free circulation assured within the Schengen area. In this evolving and sometimes antithetical landscape with both open and closed borders, our conception about the meaning and function of borders may come under challenge. This however does not mean that we must abandon the attempt to study them. In contrast, delving into their exploration becomes more pressing and urgent so as to understand their past, present, and maybe future development.

Yet, to study borders and border control in Europe, one needs to move beyond the confines of formally established rules for border control. Even if these rules do not solely include more than 3,000 pages of the Schengen acquis (Marenin, 2010: 10) but also the temporary and therefore exceptional border control measures taken in response to internal security and public policy or public health threats. Instead, we need to turn to the dynamics of constructing borders and border control, accounting therefore for the frequently overlooked meanings, routines, and actors shaping and framing the border control conduct. This enables reorienting the discussion about borders and border control in Europe towards a non-static orientation that takes into account the possibility of change, whilst studying how borders are being shaped and by whom.

So, who guards the EU external borders? Who determines how to guard them? How important is border control for Europe and its member states? Inspired by these questions and intrigued by the EU's chronic fixation with border control, this book discusses Frontex, borders, and border control. In particular, it explores Frontex's role in EU border control and at the EU external borders, delving into Frontex as a border control actor. The European Border and Coast Guard Agency, or Frontex, for short, is the EU border control agency established in 2004 'with a view to improving the integrated management of the external borders of the member states of the European Union' (Regulation, 2004). It has been operational since 2005, marking 17 years of non-stop presence at the EU external borders. Ultimately, Frontex has become inextricably linked with borders and border control in Europe. Significant growth in personnel and budget as well as a tremendous expansion in tasks and powers after continuous mandate enhancements have rendered Frontex more powerful than ever. With diverse and continuous operational actions on land, at sea, and in the air borders, at the EU territory and beyond, Frontex has drawn a different image for borders and border control. It has become 'a symbol for the European Union' (European Commission, 2016b) and 'a cornerstone of the EU's efforts to guarantee an area of freedom, security, and justice' (Frontex, 2017). All these indicate that Frontex's role surpasses the 'EU border control agency' title written on the paper of its 'business card'. From an agency, it has evolved into a border control agent shaping the EU borders.

The book's intent is to show that Frontex is not just an EU Home Affairs agency or 'more of the same' (Wolff & Schout, 2013), as the academic literature often contends. Rather, it constitutes a key border control actor that shapes borders and border control in Europe, fundamentally altering the function and notion of the 'Schengenland' area (Walters, 2002).

Border evolution

For a long time, borders were perceived as concrete boundary lines drawn between adjacent states to reflect physical features, the outcome of wars, or the result of negotiations finalised with international treaties (Agnew, 2008: 3). With the end of the 30 Years War and the signing of the Westphalian Treaty in 1648 that understanding got even more pronounced as borders and the Westphalian state became interlinked and co-defined. For the Westphalian state-centric system, a sovereign state was recognised as a territorial space clearly demarcated by concrete borders. Even maps became adjusted accordingly so as to reflect the new vision of sovereign states enacted with the Westphalian order. More specifically, prior to Westphalia most political maps were uncoloured with no representation of clear borders. Yet, in the post-Westphalian era, state borders started to be delineated with variant colours so as to clearly mark where one country ends and the other begins (Pickering, 2013). According to the Westphalian spirit, borders were considered as a symbol of sovereignty (Krasner, 1999: 3–42), namely the state's monopoly of authority over a territory (Philpott, 2001: 16–17). They belonged to a sole category, that is, territorial borders, and were conceived as static ontological entities (Sendhardt, 2013: 25) and vertical barriers separating 'us' from 'them', namely the outsiders. In this sense, they were marking the territorial integrity of states functioning as physical territorial barriers between countries or the 'frontier', that is, the front line where one meets the enemy (Anderson, 1996: 9).

However, the fall of the Berlin Wall and the era of globalisation that followed, combined with schemes of transnational integration and mobility as well as trade acceleration, heightened borders' permeability (Smith et al., 1999; Sendhardt, 2013: 24). In this increasing interconnected environment, borders were becoming less relevant (Brunet-Jailly, 2005), whereas states less powerful, as they were no longer able to assert absolute control over the flow of people, goods, capital, and ideas across borders (Cohen, 2001: 80). As a result, embedded conceptions on borders' function became seriously defied, condemning the Westphalian sovereign state to obsolescence as sovereignty did not equate any more with statehood (Shaw, 2000: 228).

In this respect, actors other than the state started to engage in borders and adopt territorial strategies (Agnew, 2009: 28), with the EU being the most prominent example (Scott, 2009). The EU project transformed the conceptualisation of borders, sparking a revolution in the understanding of sovereignty (Philpott, 2001: 39). Moving away from Westphalia's constitutive principle of sovereign states, it advocated for a post-modern and post-national narrative regarding the role of territories (Ruggie, 1993). Actually, the EU proposed a model of integration and interaction that transcends state borders by abolishing internal borders and creating new centres of governance other than the state (Lavdas & Chryssochoou, 2011: 29) as well as new political orders other than those demarcated by linear territorial boundaries (Browning, 2005).

Yet a 'borderless world' (Ohmae, 1990) did not emerge, despite the bold projections triggered by globalisation and EU integration. On the contrary, new border 'fortresses' were erected to tackle security threats, especially after the 9/11 terrorist attacks in the United States of America (USA) (2001) and later on the EU soil (2004 Madrid and 2005 London bombings). In this vein, a connection was made between terrorism, migration, and border security that led to the strengthening of border control measures to prevent organised crime and terrorist acts. Following this, border control, namely the adopted measures to regulate and monitor the borders, became a key term describing the main function of borders, especially in the USA (Andreas, 2003: 1–2).

This evolution in border conception attests borders' complexity and conflicting character, as they compose a set of opposites. They serve as barriers separating but they can also serve as bridges connecting different people. Hence, they are both open and closed, namely gates and walls. On that account, borders are way more than lines drawn on political maps (Newman, 2006: 175). They are constantly present and never cease to multiply by being continuously produced, reproduced, and transformed (Paasi, 2012: 2305). So, borders still matter and exist.

New 'kids' on the border control block

In Europe, the abolition of internal borders resulted in the simultaneous toughening of border controls outside the Schengen territory (Bigo, 2002; Huysmans, 2006) and the creation of a new border conception, that is, the EU external border. In practice, the EU external border was created by the Treaty of Amsterdam (Castan Pinos, 2009: 10–11) that was signed in 1997. Today, this border consists of almost 9,000 kilometres (km) of land borders, 42,000 km of sea borders, approximately 300 international airports (Frontex, 2022), and encloses a population of more than 440 million (Eurostat, 2021). Despite the construction of a common border both emanating from and fostering the EU integration, the EU external border also functions as a spatial line of exclusion separating the insiders from the outsiders. In fact, to keep outsiders out, the EU external border has become the deadliest border on earth (Ferrer-Gallardo & van Houtum, 2014: 297), as the adopted border control measures against irregular migration have provoked a massive death toll.

Border transformation, border control resurgence, and cross-border mobility had shifted the discourse on border matters. Borders and border control started to be recognised as 'being made' (van Houtum et al., 2005; Newman, 2011), creating new entries in the border lexicon, like 'bordering', 'b/ordering', 're-bordering', and 'de-bordering'. As a result, the ontological question of *what* is a border became replaced by the query of *who* constructs them and *how* (Sendhardt, 2013: 25). This, unavoidably, directs the attention to border control actors who act at the borders. Moreover, over the years, new non-state actors became involved in border control activities (Guiraudon & Lahav, 2000: 176; Tholen, 2010: 265; Rumford, 2012: 897; Vaughan-Williams, 2015: 6). The management and control of borders entrusted upon these actors confers them also the ability, and maybe even right, to decide who belongs inside the territory, who can enter, and who will remain or deserves to remain outside of it (Della Sala, 2017: 545). So, their role is far from technical or marginal. The same applies to Frontex, being a non-state actor engaged with border control.

Setting Frontex's scene

Frontex, undoubtedly, is a highly visible agency that, over the years, has been without cease under the media spotlight, gaining public attention, usually in the form of criticism, due to concerns on human rights breaches during its operations or institutional malpractices. This criticism, rather often, has been transfigured into the initiation of detailed investigations on Frontex's operational conduct from journalists, non-governmental organisations (NGOs), parliamentary committees, EU bodies, and courts.[1] By the same token, the academic attention on Frontex is centred on human rights considerations or accountability preoccupations (Fisher-Lescano et al., 2009; Pollak & Slominski, 2009; Baldaccini, 2010; Papastavridis, 2010; Campesi, 2014; Marin, 2014; Carrera & den Hertog, 2015; Mitsilegas, 2015; Pallister-Wilkins, 2015; Mungianu, 2016; Santos Vara, 2016; Poméon, 2017; Fink, 2018, 2020; Perkowski, 2018).

This human rights fixation from the public, the media, and academia is by far warranted and stems from Frontex's engagement with the sensitive field of border control. Being present at the border and participating in border control results in frequent interaction with vulnerable groups encountered at the borders, such as asylum seekers, minors, and victims of human smuggling. So, it is possible of any malpractice or omission to equate to a human rights infringement of the most vulnerable. Noteworthy is that Frontex, as an EU body, by virtue of its establishment and operation, has been destined to protect and promote the EU fundamental rights and core values. So, any allegation of human rights violation may seem, at least, 'peculiar' (Carrera et al., 2013), or, at most, endangers the EU's construction and projection as a 'normative power' (Manners, 2002). Yet, despite these human rights considerations, uttered almost from the start of the agency's establishment, Frontex continues enhancing its power. Actually, the battle cry of its opponents to 'shut down Frontex' appears today more utopian than ever. With an ever-expanding institutional mandate, additional powers, and newly hired staff, Frontex seems to be acting all the more self-assured. Thus, Frontex's operation and presence at the borders does not seem likely to end any time soon. Put differently, Frontex is here to stay, with no plan for its dissolution.

In this spirit, having already celebrated 17 years of continuous operations at the EU external borders and with many more to come, Frontex needs to be re-examined through a different lens. Human rights should not be the sole focus of Frontex's literature. Otherwise, important aspects of Frontex's function are in danger of being omitted or remaining hidden. As a remedy, this book proposes exploring Frontex's role in EU border control by adopting a cultural prism. The appeal to culture is not coincidental. It allows capturing borders' evolving and complex character, as both culture and borders are not fixed. They can change over time. So, culture allows accounting for any change at the EU borders. Moreover, bringing culture into Frontex's analysis permits developing a variant and original frame for Frontex's investigation that acknowledges Frontex as a member of a border control policy community. Meanwhile, prior analyses on Frontex fail to recognise the existence of a border control community with Frontex being its active member. Yet, being a border control actor, practising border control, and communicating with other border control actors, Frontex has developed into a member of a wider border control community that operates at the borders, interacting with the other border control actors. Therefore, using culture we can delve into shared meanings and common routines developed among the members of the border control community, recognising in turn the community's existence and way of interaction. Also, the absence of culture as an analytical vantage point in previous studies scrutinising this EU agency, enables this contribution to offer a distinct perspective about Frontex's role in EU border control. So, this book studies and perceives Frontex differently; as an actor of border control and member of a border control community, and not just a border control instrument or vehicle (Reid-Henry, 2013: 200) for the implementation of EU policies.

Border control actors

Whoever has crossed a border has witnessed different actors operating and interacting there, as borders function also as a place of conduct and 'topos' for social relations. Yet, being at the border does not necessarily equate to a role in border control, and, as a result, does not infer the 'border control actor' attribute. There are actors who, whilst they act at the border, do not participate in the border control conduct. For instance, despite border crossers being subject to border control, they are not its actors, because they do not implement it. Nor do they have the capacity to confer a conscious alteration to it. The same applies to NGOs that are present at the borders, like Amnesty International and the European Council on Refugees and Exiles (ECRE). Though they can formulate general suggestions and policy recommendations or criticise and put pressure on national authorities regarding border control aspects and human rights, they neither practise nor plan border control. In principle, this means that they are not and cannot be part of a border control policy community, which exchanges views and shares habits for the border control conduct. Rather, they remain external actors, which can only play a restricted role in informing the broad discussion on border control (Zaiotti, 2011: 32). For instance, a dramatic change in the number of border crossers has the potential to alter the way of carrying out border control or provide solid proof for an amendment in border policy. Yet, this does not constitute an internal alteration of the regime. Rather, it describes certain circumstances that the members of a border control policy community can acknowledge, disregard, or use for their benefit.

Instead, border control actors can be listed as those actors that conduct, practise, and implement border control at the EU external borders. In this category belong border guards from the national authorities of member states and Frontex, taking into account that the responsibility for the management of the EU external borders is shared between them (Regulation, 2019).

Obviously, apart from these actors, there are also other entities that inform border control at both national and the EU level; for instance, national parliaments that adopt border control legislation. However, this concerns a general function of the legislative procedure and not an active and conscious presence at the external borders. Similarly, at the EU level, the Directorate-General for Migration and Home Affairs (DG Home) from the side of the European Commission, the European Parliament, the Council, and the EU Court of Justice can all have an impact on the EU border control policy. However, their impact is directed towards policy-making, namely the formulation of policy and the adoption of general guidelines or legislation. Due to the technical character of the border control field, any specification of the measures applied for the border control implementation on the field cannot be a matter for reflection or decision, at the Council, the European Parliament, or even the European Commission, as these institutions do not possess the required technical expertise or the 'highly technical know-how' (Commission, 2003). For this reason, after all, Frontex was created in the first place, namely to deal with these technical aspects of border management at the EU level (Léonard, 2009). Likewise, at the national level, the actual conduct of border control is undertaken by national authorities that implement the border control policy, such as the Police, the Border Police, the Coastguard, the State Border Guard Service, the Border Guard or Civil Guard, and particularly the border guards that manage the borders, though the State as an institution exercises power in the border control field.

Therefore, the border guards and Frontex are the sole border control actors which operate at the borders, carrying out border control and participating in its conduct. Despite not being given free rein, they act together trying to confront any border control challenges erupting at the borders. This means that they actively interact and cooperate on border control-related matters. They communicate, have similar routines, and share their experiences on the field. Actually, together, Frontex and national border guards form a community of border control practitioners, which is embedded in a concrete cultural setting, as the following pages show.

Speaking of culture

We all know the word 'culture'. We frequently use it in our lexicon and everyday discussions. Beyond this, culture is a key analytical variable. It is present everywhere and shaping everything. It informs social reality (Weber, 1949, 2012 [1905]) and suspends humans in its webs of significance (Geertz, 1973). It determines what to want, to prefer, to desire, and to value (Hudson, 1997: 8). It surrounds and, in parallel, unites, and navigates actors (Gray, 1999a: 138) so as to make sense of the world and their place in it. Thus, the social world is being constituted by culture (Lapid & Kratochwil, 1994; Weldes et al., 1999). Therefore, any analysis of the social realm disregarding culture seems lacking.

Not surprisingly, after anthropology, sociology, and psychology (Boas, 1966 [1940]; Bourdieu, 1993), culture spilled over to international relations, EU studies, and security (Huntington, 1993; Lapid & Kratochwil, 1994; Katzenstein, 1996; Meyer, 2006; Williams, 2007; Lebow, 2008). Soon border studies followed with culture converting into a tool for the study of borderland communities (Wilson & Donnan, 2005; Konrad & Nicol, 2011), symbolic borders (Kurki, 2014), and us/them relationships (Newman, 2003; Dimitrovova, 2010). These reflections emphasise cross-cultural exchanges among different ethnic groups that live at the borders. Yet, besides borders, culture can also inform border control. It can exist and shape border control leading in turn to its evolution (Zaiotti, 2011).

In this spirit, the choice to apply culture into Frontex's analysis may not seem so strange after all, due to the agency's link to both borders and border control. At the same time, culture allows us to adopt a non-traditional conception by moving to the non-state level and acknowledging the role of non-state actors (Norheim-Martinsen, 2011: 528). In this light, this book explores the role of Frontex, that is, a non-state actor, in the EU border control, that is, a non-state policy. With culture we can study both the structure, namely the border, and the agency, that is, Frontex as an actor. Besides, advancing a non-traditional prism, culture enables us to turn to the ideational context, yet without disregarding the material world. This means that the insertion of culture opens up the research to shared meanings and collective understandings. So, with the culture we can put forward a different set of questions and adopt a variant research disposition that sheds light on Frontex as an actor that acts in its social environment, producing and reproducing behavioural patterns and meanings. Speaking of social environment and shared meanings, groups or communities come to mind.[2] So, we can also trace and explore the border control community, remedying another gap in the literature.

Accordingly, culture refers to collective types of behaviour and action shared among a group or community of people. In other words, everything a group does is a manifestation of culture or, it is, at least, culturally shaped (Gray, 1999b: 52), as its members share certain collective cultural traits. This means that culture can be transmitted from one individual or group to another. This transmission involves the transfer or inheritance of ideas, beliefs, values, and practices, like genes (Florini, 1996) and it can resemble to the process of learning through interaction with the physical or social environment via socialisation with others (Wendt, 1999: 123).

Apart from being transmitted through socialisation (Berger, 1996: 326), culture also enables socialisation. Indeed, culture can become a source of socialisation, as it provides the space to create relations, along with the content of communication, and the mode to interpret the others. Put differently, culture can also form and weave together groups and communities. All these denote that culture is not static but dynamic. It can change over time (Gray, 1999b: 52) as well as evolve from one culture to another.

Aside from being internalised and habiting in the minds, culture is also expressed and manifested externally, via socialisation and communication. In this sense, it can be found in words, ideas, customs, norms, beliefs, rituals, and images (Kim, 2009: 6), which permit to mark out others that have similar cultural traits. This turns the attention to the members of the group or community that share the same culture. By expressing, materialising, and pursuing a distinct culture, these members make their worlds intelligible and manageable (McNamara, 2015: 27), as they attribute meaning to the social world. In fact, through the creation and reproduction of signification signs, such as symbols, language, and practices, they interpret the world (Wedeen, 2002: 720). In this context, these members participate in the ongoing meaning-making process. They are not just 'puppets' of social structure. They define, produce, reproduce, interpret, apply, and use culture. This occurs with the adoption of practices, the articulation of arguments and ideas, as well as the shaping of everyday realities and routines.

Yet, these members of the group or community, though embedded in the cultural setting, do not always conform to established cultural patterns. The adoption and internalisation of meanings and norms, apart from being a dynamic process, can also entail variability, as the interpretation, endorsement, and diffusion of culture among the members of the community may differ, triggering, in turn, culture's gradual evolution. Likewise, the members of the community may start adopting new ideas and practices, even via their interaction with other actors, contesting therefore established cultural truths. In this spirit, the members of the group or community can alter culture by initiating new ways of acting or organising (Scott, 2008: 78), thus impacting on the social world. So, they constitute both agents and actors of culture.

As such, taking into account that no one is free of cultural traits (Herskovits & Willey, 1923: 192) and can operate beyond culture, even institutions (Gray, 1999a: 129; Scott, 2008) or organisations (Kier, 1997; Crawford, 2002), as well as that there is no aspect of life that it is not affected by culture (Hall, 1976: 16), then culture is far from ill-suited for Frontex's scrutiny. After all, every actor's behaviour is a reflection and expression of culture — and so is Frontex's.

Cultures of border control

To insert culture into Frontex's analysis, this book applies the 'cultures of border control' analytical framework. 'Cultures of border control' is a cultural evolutionary framework developed by Ruben Zaiotti (2011) with the aim to explain the institutionalisation of Schengen as a regional border regime for the management of Europe's borders. So, this framework enables not only the operationalisation of culture, but, at the same time, it constitutes a 'smooth' way of bringing culture into the EU border control field, as it was created to address the evolution of the border control culture in Europe by examining Schengen's emergence. In parallel, with its application, this book aims at reviving this framework by using it to rescrutinise Europe's borders and Schengen, more than 10 years after its conception and application by Zaiotti.

Drawing on Zaiotti (2011: 23), the culture of border control is defined as 'a relatively stable constellation of background assumptions and corresponding practices shared by a border control policy community in a given period and geographical location'. Delineating the definition's components, a border control community is consisted of a group of actors that 'share similar background assumptions and participate in common practices in the border control domain' (Zaiotti, 2011: 25). The members of the group have frequent interaction and social activities that enable them to communicate, exchange views, and develop a sensu communis or common sense. This can gradually create a sense of we-ness or, put differently, a conception of 'self' and, as a result, an image of 'the other' (Oren, 1995: 154).

Background assumptions are 'inter-subjective cognitive structures that members of the border control community rely on to interpret the reality in which they are inserted and to act upon it accordingly' (Zaiotti, 2011: 23). Empirically, they can be traced in the everyday routines and practices of the members of the community and function as a nescient compass for their actions and aspirations. They determine what actors should do, how they should act, and who they represent. Likewise, they connote which are the members of the border control community.

Practices can be conceived as arrays of organised activities in a given domain (Schatzki, 2001: 2; Zaiotti, 2011: 23). These activities can have various forms. They can be verbal, like discourses, and non-verbal, like routines. Overall, they are what members of a community do during their interactions.[3] Thus, practices are an identifiable, concrete, and relatively stable pattern of social activities over time (Zaiotti, 2011: 24).

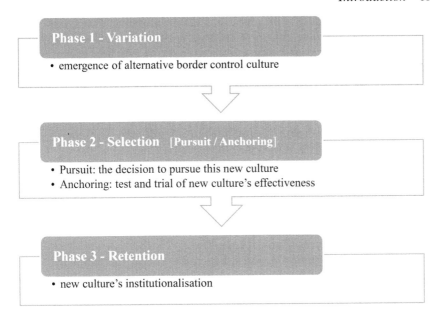

Figure 1.1 The cultural evolution process.
Source: Author's elaboration based on Zaiotti's model, 2011.

By this definition, culture is being shaped by a dynamic interplay between assumptions and practices. While assumptions define the relevant practices characterising a given domain, it is through these very practices that background assumptions are being reproduced and sustained over time, securing culture's survival and diffusion (Zaiotti, 2011: 24). But, being continuously produced and reproduced, any given culture of border control is always evolving (Zaiotti, 2011: 27). In this light, a fundamental change can lead to the culture's demise and substitution with an alternative culture. The transition from one culture to another is called 'cultural evolution' and involves different steps (see Figure 1.1; Zaiotti, 2011: 27–43).

First, a cultural variation. The variation takes place when a border control regime is considered as ineffective or seems unable to address successfully the current border control challenges. Then, the members of the border control community, after having started questioning the culture's relevance, deliberate on alternative solutions. This leads to different assumptions and practices about borders and border control, and therefore to the articulation of new measures and policies that have the potential to form an alternative border control culture challenging the dominant culture.

The second step is culture's selection, which includes two phases: culture's pursuit and culture's anchoring. The pursuit of a new culture presupposes a considerable change within the border control community, which now deems rational to pursue a new culture, in order to find an antidote to the regime's failings (Zaiotti, 2011: 34). So, culture's pursuit refers to the decision to pursue this new culture. Conversely, culture's anchoring involves culture's performance and actual effectiveness in tackling the practical challenges that preoccupy the policy domain. After all, it is the effectiveness of a culture which persuades the members of the border control policy community to adopt it. In this context, the activities carried out by a community in the beginning of a culture's pursuit can be regarded as 'experiments' that test and trial the culture, providing hard proof before its final jurisdiction, namely its acceptance or rejection (Zaiotti, 2011: 35).

The third and last step for a cultural evolution is culture's retention. Retention takes place with the institutionalisation of the new culture as the dominant culture of border control through the integration of its key tenets in official documents. Otherwise, if the alternative culture is unsuccessful, it can die or lie down, with a view to being resumed later given a favourable change in circumstances (Zaiotti, 2011: 37).

In the cultural evolution process, key is the role of the members of the community, as, apart from structuring the debate around the culture's main elements, they also apply, reproduce, and diffuse its tenets, testing and practising this new culture, before adopting and internalising it. So, the members of the community hold culture's configuration and future trajectory in their hands.

To Schengen or not to Schengen?

Based on the cultural evolutionary process, conceived by Zaiotti (2011: 45–89), three distinct cultures of border control in Europe are identified: 'Westphalia', 'Schengen', and 'Brussels'. Westphalia characterises a nationalist approach to border control with strict measures, according to which borders function as barriers separating states. This approach was formed in the beginning of the modern state system and is based on the principles of the 1648 Treaty of Westphalia emphasising territorial integrity and ethnic homogeneity of the sovereign state. It was the dominant culture in Europe, mostly after World War II and until the 1980s (Zaiotti, 2011: 47). This period was characterised by severe obstacles to movement and limited cross-border flows.

However, after reaching its full maturation, the Westphalia culture of border control started to be seriously contested. The fall of the Iron Curtain in 1989, coupled with the deepening of European integration as well as intense transnational mobility, due to globalisation, unavoidably impacted the conception of borders, shaking the nationalist regime of Westphalia. In that period, many parts of the European continent started to witness the results of the economic recession (Zaiotti, 2011: 61–62). The recession caused high levels of immigration within Europe. In parallel, it brought migratory flows from and to Europe with the arrival of a large number of non-Europeans. This high scale of migration, along with the rise of terrorism in many

European countries in the 1970s and 1980s, started to raise concerns about the nature of borders, linking border control with security (Zaiotti, 2011: 64–65). Hence, in practice, although the international and European context facilitated the lifting of economic and trade barriers, it did not result in the free movement of people. On the contrary, it triggered a security anxiety and a re-invigoration of borders.

From the second part of the 1980s until the late 1990s, two alternative cultures of border control emerged in Europe: Schengen and Brussels. Despite being different, these two cultures had the same goal, that is, a Europe without borders, contesting therefore the Westphalian cultural trajectory.

Schengen took its name from the intergovernmental 1985 Schengen Agreement signed by Germany, France, the Netherlands, Belgium, and Luxembourg on the gradual abolition of checks at their common borders. This text embodies a novel approach to border control — a transgovernmental approach — which sets a distinction between external and internal borders. This distinction refers to a free circulation within Schengen after the abolition of internal borders and a parallel hardening of security provisions for entry at the external border to address any security vacuum.[4]

Brussels constitutes an alternative approach promoted mainly by the European Commission, aiming at a Europe without internal borders, fulfilling the spirit of its 1985 White Paper for the completion of the internal market, the 1986 Single European Act, and Maastricht Treaty. According to the Brussels model, borders have a negative connotation. They are barriers that hinder transactions. So, they should be eliminated or start functioning positively as bridges or symbols of a new collective European identity (Zaiotti, 2011: 81). In this regard, border control acquires supranational characteristics with a balanced distribution of responsibility among member states, due to the development of the notion of 'Community's borders', whereas the internal/external dimension seems irrelevant.

Schengen and Brussels developed for some years in parallel. The fact that they had the same goal of abolishing Europe's internal borders allowed them not to be antithetical and rival each other (Zaiotti, 2011: 88). Instead, both tried to contest the still-dominant Westphalia regime. Notwithstanding, only one could be anointed Europe's dominant border control culture. Accordingly, solely Schengen's attempt was deemed successful. Schengen's flexibility, positive practical results, and membership expansion resulted in the progressive internalisation of the Schengen border control culture by the border control community (Zaiotti, 2011: 160) and the regime's formal incorporation, as the Schengen acquis, into the European Union. Therefore, Schengen, after replacing Westphalia, became the new official approach to border control in Europe (Zaiotti, 2011: 144) and the dominant culture of border control from the 1990s until maybe today.

But, is Schengen still the dominant culture of border control? This question will be answered in the following pages. Applying the 'cultures of border control' framework and zooming into Frontex as a policy actor of border control allows this research to detect and then reveal any border control evolution of the Schengen regime.

After all, neither borders, border control, nor the members of the border control community remain frozen across time. Instead, they evolve, adapting to the imperatives of their social environment. This means that borders and border control may have been subjected to detrimental changes since the Schengen regime became settled and institutionalised in the 1990s and 2000s. We may even have a new border control approach or dominant culture for border control shaping Europe's borders.

A new border control culture?

Over the pages that follow, this book unfolds Frontex's role in EU border control. Applying the 'cultures of border control' framework, it explores whether and how Frontex impacts the EU border control. In particular, it scrutinises how Frontex shapes the culture of EU border control. Its aim is to elucidate both the culture of EU border control and Frontex's impact, putting forward an original theoretical context and raw material drawn from primary empirical research at the borders.

The argument advanced is that Frontex shapes the EU border control. Frontex's impact on EU border control is demarcated and materialised through culture. Accordingly, Frontex has produced and is promoting a new border control culture, which I have decided to label as 'Warsaw', due to Frontex's central involvement. Frontex's promotion of Warsaw culture shows that Frontex, from a tool or a policy instrument, has become an actor. An actor that is capable of producing new meanings and actions as well as diffusing its own assumptions and practices for border control, that is, a new culture. In fact, with Frontex's deeds, Warsaw culture has now become the dominant culture of border control replacing the Schengen regime. That replacement conveys different assumptions and practices for border control extracted from the research as well as a different border control community being formed, denoting thereby culture's evolution from Schengen to Warsaw. Thus, Frontex shapes the EU border control by constructing and pursuing Warsaw culture.

Regardless of any challenges at the borders, triggered, for example, by Brexit, COVID-19 pandemic, and 'hard' security preoccupations in the context of war or war threats in Europe, Warsaw culture savours its dominant position in EU border control, still remaining uncontested by potential alternative cultural trajectories. The argument and research result gains added importance taking into account the particular fluid and therefore uncertain context for border control, dominated by escalating migratory pressures, emerging challenges, and, in turn, increased public attention regarding Frontex's doings.

Introduction 15

The empirical research involves direct observation and in situ analysis with the conduct of fieldwork in 2018 at the Greek land border Evros and the Italian sea border Lampedusa, as well as 13 face-to-face semi-structured interviews with border control practitioners (3 with Frontex officers in Warsaw, 6 with Greek border officers in Evros and Athens, 4 with Italian border officers in Lampedusa). All interviews were conducted in Greek, Italian, and English language without the use of translators in order to avoid the involvement of any third party in the communicative, interactive, and meaning-making action of interviewing. My past work at Frontex proved to be rather valuable in tracing and approaching prospective interviewees. Equally important was my decision to contact prospective interviewees directly, by email or phone, and ask them to participate in my research or through 'snowballing', especially upon encountering characteristically delayed answers or negative responses from the institutions' gatekeepers. Other research methods used in the subsequent chapters include document analysis of primary and secondary sources, institutional discourse analysis, mainly of Frontex and local border control services' narratives, process-tracing, and comparative case study analysis of Evros and Lampedusa borders.

The selection of these two borders has not been random. Rather, they constitute two symbolic EU external borders with distinct characteristics. On the one hand, Evros is the external border of the EU with Turkey in northern Greece along the river Evros, which, for many years, has been the main point of entry for irregular migrants to reach Europe. On the other hand, Lampedusa constitutes Italy's most southern part, an island close to Tunisia and Libya, which in the last years has also become a point of access for irregular migrants trying to enter Europe. Thus, Evros and Lampedusa are two borders that are different in geopolitical, symbolic, functional, and geographic terms. The only common element that they share as EU external borders is that both have attracted vast migration pressures and, as a result, have functioned as the operational theatre for various Frontex joint operations and activities. Hence, it is possible to draw a wider conclusion about common patterns observed in different cases (Mill, 1865 [1843]), such as the border control conduct and Frontex's role at Europe's borders. Any conclusions drawn from this comparison will be further elaborated by adding in the analysis a newly established border, that, is the United Kingdom (UK)-France border, which after Brexit has evolved into an EU external border.

Book overview

This introductory part sought to set out the context of the book. The next chapter focuses on Frontex (Chapter 2). Analysing its function, tasks, and structure, I emphasise Frontex's border control aspects. Moreover, reviewing the literature on Frontex's role in EU border control, I highlight the main themes monopolising the discussion and the lacunae that this book seeks to address.

In Chapter 3, I insert the 'cultures of border control' framework into Frontex's analysis. The chapter is divided into two sections. The first part looks at Frontex's role as a border control actor and its central position in the border control community. The second part presents Frontex's border control assumptions and practices extracted via document analysis, institutional discourse analysis, and interviews with Frontex staff.

Chapters 4 and 5 examine both the border and border control conduct. They include the case study analysis of the two borders, namely Lampedusa (Chapter 4) and Evros (Chapter 5). Drawing data mainly from fieldwork and interviews with national border guards in Evros and Lampedusa, these two chapters elucidate the assumptions and practices for the border control conduct, uncovering common elements that, due to the variant nature of these two borders, constitute cultural traits.

In Chapter 6, firstly, I synthesise the previous chapters' research results. The synthesis shows common border control assumptions and practices traced both in different border locations (Lampedusa and Evros) and within a border control actor (Frontex). Yet, the assumptions and practices extracted vary from the Schengen regime. This indicates the emergence of a new border control culture, which I label as 'Warsaw' border control culture. In the next part, I compare Warsaw culture with the other border control approaches, as described in the analytical framework of the 'cultures of border control', namely Schengen, Brussels, and Westphalia. This comparison shows that Warsaw culture constitutes a new border control trajectory, which, after contesting Schengen, now constitutes the dominant culture for border control. The last sections assess Frontex's impact on the development and evolution of Warsaw culture as well as discuss the role of EU institutions in Warsaw's consolidation. Following a sequential process-tracing path and extracting data from document analysis, institutional discourse analysis, interviews, and process-tracing evidence, the chapter reveals that Frontex impacts on the culture of EU border control, and therefore shapes EU border control, by producing and promoting the components of the Warsaw border control culture.

With Chapter 7, I reflect on major challenges ahead facing the border control policy community and Warsaw culture. The first part explores the transformation of the English Channel from internal to external border. The next part refers to border and border control changes triggered by the COVID-19 pandemic. The last part deals with 'hard' security matters, discussing firstly the escalation of tensions between Greece and Turkey and secondly the recent crisis in Ukraine after Russian invasion.

Chapter 8 concludes by summarising the key findings and main conclusions drawn from the preceding chapters. It also discusses the new knowledge produced by this book regarding Frontex, borders, and border control in Europe. Finally, it reflects on wider issues stemming from agency governance and cultural inter-links.

Notes

1 For instance, the European Anti-Fraud Office (OLAF), the European Agency for Fundamental Rights (FRA), the European Ombudsman, the European Court of Justice (CJEU), the European Parliament's Committee on Civil Liberties, Justice, and Home Affairs (LIBE), which established a Working Group on Frontex Scrutiny, and NGOs, like the Human Rights Watch.
2 For the interplay between ideas and actors, see also Blyth (2002) and Saurugger (2013).
3 Practices at the collective level differ from those at the individual level, which can be characterised as habits. Rather, shared practices are a collective accomplishment (Barnes, 2001: 23).
4 Today, the Schengen area is comprised of 26 countries; 22 EU countries; and 4 European Free Trade Association states, namely Iceland, Liechtenstein, Norway, and Switzerland.

References

Agnew, J. (2008) 'Borders on the Mind: Re-Framing Border Thinking'. *Ethics & Global Politics*, 1(4), pp. 1–17.
Agnew, J. (2009) *Globalization and Sovereignty*. Lanham: Rowman & Littlefield.
Anderson, M. (1996) *Frontiers: Territory and State Formation in the Modern World*. Cambridge: Polity Press.
Andreas, P. (2003) 'A Tale of Two Borders: The US-Mexico and US-Canada Lines after 9/11'. In: P. Andreas & T.J. Biersteker (eds.), *The Rebordering of North America: Integration and Exclusion in a New Security Context*. London: Routledge, pp. 1–23.
Baldaccini, A. (2010) 'Extraterritorial Border Controls in the EU: The Role of Frontex in Operations at Sea'. In: B. Ryan & V. Mitsilegas (eds.), *Extraterritorial Immigration Control: Legal Challenges*. Leiden: Martinus Nijhoff, pp. 229–256.
Barnes, B. (2001) 'Practice as Collective Action'. In: T.R. Schatzki, K. Knorr-Cetina & E. von Savigny (eds.), *The Practice Turn in Contemporary Theory*. Abingdon: Routledge, pp. 17–28.
Berger, T.U. (1996) 'Norms, Identity and National Security in Germany and Japan'. In: P.J. Katzenstein (ed.), *The Culture of National Security: Norms and Identity in World Politics*. New York: Columbia University Press, pp. 317–356.
Bigo, D. (2002) 'Security and Immigration: Toward a Critique of the Governmentality of Unease'. *Alternatives*, 27(1), pp. 63–92.
Blyth, M. (2002) *The Great Transformation: Economic Ideas and Institutional Change in the 20th Century*. Cambridge: Cambridge University Press.
Boas, F. (1966 [1940]) *Race, Language and Culture*. New York: The Free Press.
Bourdieu, P. (1993) *The Field of Cultural Production*. Cambridge: Polity Press.
Browning, C. (2005) 'Westphalian, Imperial, Neo-mediaeval: The Geopolitics of Europe and the Role of the North'. In: C. Browning (ed.), *Remaking Europe in the Margins: Northern Europe after the Enlargements*. Aldershot: Ashgate, pp. 85–101.
Brunet-Jailly, E. (2005) 'Theorizing Borders: An Interdisciplinary Perspective'. *Geopolitics*, 10(4), pp. 633–649.

Campesi, G. (2014) 'Frontex, the Euro-Mediterranean Border and the Paradoxes of Humanitarian Rhetoric'. *South East European Journal of Political Science*, II(3), pp. 126–134.

Carrera, S. & den Hertog, L. (2015) 'Whose Mare? Rule of Law Challenges in the Field of European Border Surveillance in the Mediterranean'. *CEPS*, Paper in Liberty and Security in Europe 79, pp. 1–29.

Carrera, S., den Hertog, L. & Parkin, J. (2013) 'The Peculiar Nature of EU Home Affairs Agencies in Migration Control: Beyond Accountability versus Autonomy?'. *European Journal of Migration and Law*, 15(4), pp. 337–358.

Castan Pinos, J. (2009) 'Building Fortress Europe? Schengen and the Cases of Ceuta and Melilla'. *Centre for International Border Research*, Working Paper 14, pp. 1–29.

Cohen, E. (2001) 'Globalization and the Boundaries of the State: A Framework for Analyzing the Changing Practice of Sovereignty'. *Governance*, 14(1), pp. 75–97.

Commission of the European Communities (2003) 'Proposal for a Council Regulation establishing a European Agency for the Management of Operational Co-Operation at the External Borders'. COM(2003)687.

Crawford, N.C. (2002) *Argument and Change in World Politics: Ethics, Decolonization and Humanitarian Intervention*. Cambridge: Cambridge University Press.

Della Sala, V. (2017) 'Homeland Security: Territorial Myths and Ontological Security in the European Union'. *Journal of European Integration*, 39(5), pp. 545–558.

Dimitrovova, B. (2010) 'Cultural Bordering and Re-Bordering in the EU's Neighbourhood: Members, Strangers or Neighbours?'. *Journal of Contemporary European Studies*, 18(4), pp. 463–481.

Eurobarometer (2019) 'Autumn 2019 Standard Eurobarometer'. Available at https://ec.europa.eu/commission/presscorner/detail/en/ip_19_6839 (accessed November 2021).

European Commission (2016a) 'State of the Union 2016: Commission Targets Stronger External Borders'. Available at https://europa.eu/rapid/press-release_IP-16-3003_en.htm (accessed January 2022).

European Commission (2016b) 'Remarks by Commissioner Avramopoulos at the Launch of the European Border and Coast Guard Agency'. Available at https://europa.eu/rapid/press-release_SPEECH-16-3334_en.htm (accessed January 2022).

European Commission (2021) 'Asylum and Return: Commission Proposes Temporary Legal and Practical Measures to Address the Emergency Situation at the EU's External Border with Belarus'. Available at https://ec.europa.eu/commission/presscorner/detail/en/IP_21_6447 (accessed May 2022).

Eurostat (2021) 'Population and Population Change Statistics'. Available at http://ec.europa.eu/eurostat/statistics-explained/index.php/Population_and_population_change_statistics (accessed November 2021).

Ferrer-Gallardo, X. & van Houtum, H. (2014) 'The Deadly EU Border Control'. *ACME: An International E-Journal for Critical Geographies*, 13(2), pp. 295–304.

Fink, M. (2020) 'The Action for Damages as a Fundamental Rights Remedy: Holding Frontex Liable'. *German Law Journal*, 21(3), pp. 532–548.

Fink, M. (2018) *Frontex and Human Rights: Responsibility in 'Multi-Actor Situations' under the ECHR and EU Public Liability Law*. Oxford: Oxford University Press.

Fisher-Lescano, A., Löhr, T. & Tohidipur, T. (2009) 'Border Controls at Sea: Requirements under International Human Rights and Refugee Law'. *International Journal of Refugee Law*, 21(2), pp. 256–296.

Florini, A. (1996) 'The Evolution of International Norms'. *International Studies Quarterly*, 40(3), pp. 363–389.
Frontex (2017) 'Frontex Celebrates Entry into Force of the Agency's Headquarters Agreement'. Available at https://frontex.europa.eu/media-centre/news/news-release/frontex-celebrates-entry-into-force-of-the-agency-s-headquarters-agreement-6ibIQc (accessed January 2022).
Frontex (2020) *Risk Analysis for 2020*. Warsaw: Risk Analysis Unit.
Frontex (2022) 'Information Management'. Available at https://frontex.europa.eu/we-know/situational-awareness-and-monitoring/information-management/ (accessed January 2022).
Geertz, C. (1973) *The Interpretation of Cultures: Selected Essays*. New York: Basic Books.
Gray, C.S. (1999a) *Modern Strategy*. Oxford: Oxford University Press.
Gray, C.S. (1999b) 'Strategic Culture as Context: The First Generation of Theory Strikes Back'. *Review of International Studies*, 25(1), pp. 49–69.
Guiraudon, V. & Lahav, G. (2000) 'A Reappraisal of the State Sovereignty Debate: The Case of Migration Control'. *Comparative Political Studies*, 33(2), pp. 163–195.
Hall, E.T. (1976) *Beyond Culture*. New York: Doubleday.
Herskovits, M.J. & Willey, M.M. (1923) 'The Cultural Approach to Sociology'. *American Journal of Sociology*, 29(2), pp. 188–199.
Hudson, V.M. (1997) 'Culture and Foreign Policy: Developing a Research Agenda'. In: M. Hudson (ed.), *Culture and Foreign Policy*. Boulder: Lynne Rienner, pp. 1–26.
Huntington, S.P. (1993) 'The Clash of Civilizations?'. *Foreign Affairs*, 72(3), pp. 22–49.
Huysmans, J. (2006) *The Politics of Insecurity: Fear, Migration and Asylum in the EU*. Abingdon: Routledge.
Katzenstein, P.J. (ed.) (1996) *The Culture of National Security: Norms and Identity in World Politics*. New York: Columbia University Press.
Kier, E. (1997) *Imagining War: French and British Military Doctrine between the Wars*. Princeton: Princeton University Press.
Kim, J. (2009) *Cultural Dimensions of Strategy and Policy*. Carlisle: Strategic Studies Institute, US Army War College.
Konrad, V. & Nicol, H.N. (2011) 'Border Culture, the Boundary between Canada and the United States of America, and the Advancement of Borderlands Theory'. *Geopolitics*, 16(1), pp. 70–90.
Krasner, S.D. (1999) *Sovereignty: Organized Hypocrisy*. Princeton: Princeton University Press.
Kurki, T. (2014) 'Borders from the Cultural Point of View: An Introduction to Writing at Borders'. *Culture Unbound*, 6(6), pp. 1055–1070.
Lapid, Y. & Kratochwil, F. (eds.) (1994) *The Return of Culture and Identity in IR Theory*. London: Lynne Rienner.
Lavdas, K. & Chryssochoou, D.N. (2011) *A Republic of Europeans: Civic Potential in a Liberal Milieu*. Cheltenham: Edward Elgar.
Lebow, R.N. (2008) *A Cultural Theory of International Relations*. Cambridge: Cambridge University Press.
Léonard, S. (2009) 'The Creation of FRONTEX and the Politics of Institutionalisation in the EU External Borders Policy'. *Journal of Contemporary European Research*, 5(3), pp. 371–388.
Manners, I. (2002) 'Normative Power Europe: A Contradiction in Terms?'. *Journal of Common Market Studies*, 40(2), pp. 235–258.

Marenin, O. (2010) 'Challenges for Integrated Border Management in the European Union'. *DCAF*, Occasional Paper 17, pp. 1–161.

Marin, L. (2014) 'Policing EU's External Borders: A Challenge for the Rule of Law and Fundamental Rights in the Area of Freedom, Security and Justice?'. *Journal of Contemporary European Research*, 7(4), pp. 468–487.

McNamara, K.R. (2015) 'Imagining Europe: The Cultural Foundations of EU Governance'. *Journal of Common Market Studies*, 53(S1), pp. 22–39.

Meyer, C.O. (2006) *The Quest for a European Strategic Culture: Changing Norms on Security and Defence in the European Union*. Hampshire: Palgrave Macmillan.

Mill, J.S. (1865 [1843]) *A System of Logic: Ratiocinative and Inductive*. London: Longmans.

Mitsilegas, V. (2015) *The Criminalisation of Migration in Europe: Challenges for Human Rights and the Rule of Law*. London: Springer.

Mungianu, R. (2016) *Frontex and Non-Refoulement: The International Responsibility of the EU*. Cambridge: Cambridge University Press.

Newman, D. (2003) 'On Borders and Powers: A Theoretical Framework'. *Journal of Borderlands Studies*, 18(1), pp. 13–25.

Newman, D. (2006) 'Borders and Bordering: Towards an Interdisciplinary Dialogue'. *European Journal of Social Theory*, 9(2), pp. 171–186.

Newman, D. (2011) 'Contemporary Research Agendas in Border Studies: An Overview'. In: D. Wastl-Water (ed.), *The Ashgate Research Companion to Border Studies*. Farnham: Ashgate, pp. 3347.

Norheim-Martinsen, P.M. (2011) 'EU Strategic Culture: When the Means Becomes the End'. *Contemporary Security Policy*, 32(3), pp. 517–534.

Ohmae, K. (1990) *The Borderless World: Power and Strategy in the Interlinked Economy*. New York: Harper Business.

Oren, I. (1995) 'The Subjectivity of the "Democratic" Peace: Changing U.S. Perceptions of Imperial Germany'. *International Security*, 20(2), pp. 147–184.

Paasi, A. (2012) 'Commentary-Border Studies Reanimated: Going beyond the Territorial/Relational Divide'. *Environment and Planning A: Economy and Space*, 44(10), pp. 2303–2309.

Pallister-Wilkins, P. (2015) 'The Humanitarian Politics of European Border Policing: Frontex and Border Police in Evros'. *International Political Sociology*, 9(1), pp. 53–69.

Papastavridis, E. (2010) 'Fortress Europe and FRONTEX: Within or Without International Law?'. *Nordic Journal of International Law*, 79(1), pp. 75–111.

Perkowski, N. (2018) 'Frontex and the Convergence of Humanitarianism, Human Rights and Security'. *Security Dialogue*, 49(6), pp. 457–475.

Philpott, D. (2001) *Revolutions in Sovereignty: How Ideas Shaped Modern International Relations*. Princeton: Princeton University Press.

Pickering, S. (2013) 'Borderlines: Maps and the Spread of the Westphalian State from Europe to Asia Part One: The European Context'. *The International Archives of the Photogrammetry, Remote Sensing and Spatial Information Sciences*, XL-4(W3), pp. 111–116.

Pollak, J. & Slominski, P. (2009) 'Experimentalist but Not Accountable Governance? The Role of Frontex in Managing the EU's External Borders'. *West European Politics*, 32(5), pp. 904–924.

Poméon, A. (2017) *FRONTEX and the EBCGA: A Question of Accountability*. Oisterwijk: Wolf Legal.

Regulation (EC) No 2007/2004 of 26 October 2004 Establishing a European Agency for the Management of Operational Cooperation at the External Borders of the Member States of the European Union [2004, OJ L 349/1].

Regulation (EU) 2019/1896 of the European Parliament and of the Council of 13 November 2019 on the European Border and Coast Guard and Repealing Regulations (EU) No 1052/2013 and (EU) 2016/1624 [2019, OJ L 295/1].

Reid-Henry, S.M. (2013) 'An Incorporating Geopolitics: Frontex and the Geopolitical Rationalities of the European Border'. *Geopolitics*, 18(1), pp. 198–224.

Ruggie, J.G. (1993) 'Territoriality and Beyond: Problematizing Modernity in International Relations'. *International Organization*, 47(1), pp. 139–174.

Rumford, C. (2012) 'Towards a Multiperspectival Study of Borders'. *Geopolitics*, 17(4), pp. 887–902.

Santos Vara, J. (2016) 'In Deep Water: Towards a Greater Commitment for Human Rights in Sea Operations Coordinated by Frontex?'. *European Journal of Migration and Law*, 18(1), pp. 65–87.

Saurugger, S. (2013) 'Constructivism and Public Policy Approaches in the EU: From Ideas to Power Games'. *Journal of European Public Policy*, 20(6), pp. 888–906.

Schatzki, T. (2001) 'Introduction: Practice Theory'. In: T.R. Schatzki, K. Knorr-Cetina & E. von Savigny (eds.), *The Practice Turn in Contemporary Theory*. Abingdon: Routledge, pp. 1–14.

Scott, J.W. (2009) 'Bordering and Ordering the European Neighbourhood. A Critical Perspective on EU Territoriality and Geopolitics'. *TRAMES: A Journal of the Humanities and Social Sciences*, 13(3), pp. 232–247.

Scott, W.R. (2008) *Institutions and Organizations: Ideas and Interests*. London: SAGE.

Sendhardt, B. (2013) 'Border Types and Bordering Processes: A Theoretical Approach to the EU/Polish-Ukrainian Border as a Multi-Dimensional Phenomenon'. In: A. Lechevalier & J. Wielgohs (eds.), *Borders and Border Regions in Europe: Changes, Challenges and Chances*. Bielefeld: Transcript, pp. 21–43.

Shaw, M. (2000) *Theory of the Global Sate: Globality as Unfinished Revolution*. Cambridge: Cambridge University Press.

Smith, D.A., Solinger, D.J. & Topik, S.C. (eds.) (1999) *States and Sovereignty in the Global Economy*. London: Routledge.

Tholen, B. (2010) 'The Changing Border: Developments and Risks in Border Control Management of Western Countries'. *International Review of Administrative Sciences*, 76(2), pp. 259–278.

van Houtum, H., Kramsch, O. & Ziefhofer, W. (eds.) (2005) *B/ordering Space*. Aldershot: Ashgate.

Vaughan-Williams, N. (2015) *Europe's Border Crisis: Biopolitical Security and Beyond*. Oxford: Oxford University Press.

Walters, W. (2002) 'Mapping Schengenland: Denaturalizing the Border'. *Environment & Planning D: Society & Space*, 20(5), pp. 561–580.

Weber, M. (1949) *The Methodology of the Social Sciences*. Glencoe: Free Press.

Weber, M. (2012 [1905]) *The Protestant Ethic and the Spirit of Capitalism*. Abingdon: Routledge.

Wedeen, L. (2002) 'Conceptualizing Culture: Possibilities for Political Science'. *American Political Science Review*, 96(4), pp. 713–728.

Weldes, J., Laffey, M., Gusterson, H. & Duvall, R. (eds.) (1999) *Cultures of Insecurity: States, Communities and the Production of Danger*. Minneapolis: University of Minnesota Press.

Wendt, A. (1999) *Social Theory of International Politics*. Cambridge: Cambridge University Press.
Williams, M.C. (2007) *Culture and Security: Symbolic Power and the Politics of International Security*. Abingdon: Routledge.
Wilson, T.M. & Donnan, H. (2005) 'Territory, Identity and the Places In-Between: Culture and Power in European Borderlands'. In: T.M. Wilson & H. Donnan (eds.), *Culture and Power at the Edges of the State: National Support and Subversion in European Border Regions*. Münster: Lit Verlag Frankfurt, pp. 1–29.
Wolff, S. & Schout, A. (2013) 'Frontex as Agency: More of the Same?'. *Perspectives on European Politics and Society*, 14(3), pp. 305–324.
Zaiotti, R. (2011) *Cultures of Border Control: Schengen & the Evolution of European Frontiers*. Chicago: University of Chicago Press.

2 Frontex
An insurgent border control actor

'Hey, we're doing the right things. It works!'.
—Ilkka Laitinen, first Frontex Executive Director
(Frontex, 2010)

Introduction

Frontex constitutes an EU institutional paradox. On the one hand, it is one of the most highly criticised EU agencies, triggering heated discussions even at the EU level about its functions, operational actions, and effectiveness. In this line, the European Court of Auditors (2021) deemed Frontex 'not sufficiently effective', due to shortfalls in the implementation of its previous mandate. Yet, on the other hand, it is also one of the most rapidly growing EU agencies, acquiring new roles, additional powers, and never-decreasing resources to implement its enhanced mandate. This paradox signalises that Frontex, apart from being controversial, also represents a unique case in the EU institutional mechanism. In other words, Frontex is not just another EU agency but a key border control actor shaping the EU borders.

Beginning the analysis with Frontex, the chapter presents an overview of this insurgent EU agency outlining its main characteristics and functions. This presentation includes the reasons that led to its establishment, its governance, its significant expansion, and main tasks. The chapter also reviews the literature that engages with Frontex's role in EU border control, identifying three main streams: the security stream, the institutionalist stream, and the border stream. The review reveals that Frontex's analysis still has certain important gaps, which this book seeks to address with the insertion of culture into the study of Frontex. Actually, the application of a cultural analytical framework allows exploring Frontex as a border control actor and member of a border control community and not treating it as a mere tool of border control.

DOI: 10.4324/9781003230250-2

Frontex's birth

Borders were always a chief preoccupation for Europeans. After continuous and some of the deadliest wars, borders, instead of dividing, set in motion the EU project. They became a stimulus for unification and integration, leading to a long-lasting peace among traditional adversaries, such as the UK and Germany or France and Spain. In this context, the EU soon became a case of successful border conflict transformation, irrevocably changing the traditional meaning and function of territoriality (Diez et al., 2006). This was spurred with the Schengen Agreement and became finalised with the decision to lift border controls inside the EU territory, establishing therefore free movement of goods, services, capital, and people.[1] Yet, this decision did not eliminate borders. Conversely, it created new divisions, other enemies, and a different border type — the EU external border.

Upon the abolition of internal border checks, member states' national borders became transformed into the EU's external borders delineating EU insiders and outsiders. From that point, the burden of EU border protection was placed on frontline countries, like Greece and Italy, which, being situated along the EU's external southern border, have traditionally functioned as major cross-border corridors for both regular and irregular mobility flows. But, becoming part of the EU external border, these countries started to be held accountable and were considered responsible for the border protection of the whole EU territory, leading to the drawing of a new border management context in the wake of responsibility asymmetries among member states and insecurity considerations stemming from this new border control reality. Evoking its shared competence in this policy sphere, conferred with the 1997 Treaty of Amsterdam, the EU started laying the foundation for a common policy on migration and border management.

In this line, the EU's Migration and Home Affairs policy area, previously referred to as the Area of Freedom, Justice, and Security or Justice and Home Affairs, was set in place. Soon it became one of the most rapidly developing EU domains (Monar, 2006: 495). Yet, the policy's intergovernmental past, deriving from the Schengen Agreement, and member states' reluctance to achieve deeper integration resulted in policy fragmentation and lacking results. A characteristic example of this fragmentation constituted the establishment, under the Council of the European Union, of more than 30 committees and working parties discussing simultaneously border control matters (Monar, 2006: 499), like the 'Strategic Committee on Immigration, Frontiers, and Asylum' (SCIFA), a gathering of senior level officials, and the SCIFA+ formation consisting of the members of SCIFA and the heads of member state border control services. Yet, this institutional set-up was deemed incapable of addressing efficiently the emerging border control imperatives, due to vast membership, large agenda, funding shortcomings, and diverging visions (House of Lords, 2003: 14; Pollak & Slominski, 2009: 908; Ekelund, 2014: 105). The negative appraisal continued as the establishment of an External

Borders Practitioners Common Unit, consisting of the heads of member state border guard services, did not provide a valid outlet (Commission, 2003a; Léonard, 2009: 378–379; Neal, 2009: 342).

To remedy institutional and operational gaps, the 2003 Thessaloniki European Council, held in a post-9/11 environment, under the pressure of increased irregular migratory flows induced by the war in Iraq as well as border control concerns emanating from the forthcoming big-bang enlargement, called for a new institutional structure that would enhance the 'operational cooperation for the management of external borders' (Commission, 2003b). Responding to this call, the European Commission proposed the creation of an EU agency, namely a distinct entity, outside the normal governmental framework, carrying out technical, scientific, or managerial tasks (Majone, 2006: 191; Groenleer, 2009: 82), such as those usually encountered in border control. At the same time, this option functioned as a compromise between the community approach of the European Commission and the European Parliament, which preferred a European Border Guard corps (Jorry, 2007: 2), and member states' reluctance to abdicate power (Colimberti, 2008; Kasparek, 2010: 123). As a result, only a few months later, in November 2004, Frontex was established by Council Regulation as the EU's 19th decentralised European regulatory agency.

Legally, it was created on the basis of the provisions of the Article 62(2) (a) and 66 of the Treaty establishing the European Community. Institutionally, being an EU agency, it constitutes an autonomous administrative entity. Furthermore, the EU agency option demonstrates the vertical transfer of powers from the national to EU level, signalling a shift from national coordination with SCIFA or the External Borders Practitioners Common Unit to a more supranational approach (Rijpma, 2009: 132). In symbolic terms, Frontex's establishment represented a major breakthrough, as it was, and still is, the highest level of integration at the EU external border (Marenin, 2010: 20).

Frontex's governance

Frontex is a specialised and independent EU agency, seated in Warsaw, the capital of Poland. Its founding aim was to improve the coordination of operational cooperation between member states in external border management (Regulation, 2004). To do so, it has been granted legal personality as well as operational, technical, administrative, legal, and financial autonomy (Regulation, 2004, 2019). In this context, it has the liberty to appropriate financial, material, and human resources as it deems fit in order to fulfil its tasks.

The agency is governed and overseen by its management board, consisting of representatives of the heads of the border authorities of the EU member states that are signatories of the Schengen acquis, plus two representatives from the European Commission.[2] The management board provides strategic direction to the agency, especially through the adoption of programming documents and multiannual strategies. Furthermore, it controls Frontex via the establishment of its budgets and the verification of their execution.

Frontex is managed and represented by its Executive Director. The Executive Director is proposed by the European Commission and appointed for a 5-year term by the management board, to which board the Executive Director is accountable. The Executive Director is responsible for the day-to-day management of the agency, acts as appointing authority on behalf of Frontex, prepares and implements work programmes and activities for consideration and adoption by the management board, and is afforded independence. Until today, this post has been filled twice: initially, by the Finnish Ilkka Laitinen, who served as the first Executive Director, then, in 2015, by the French Fabrice Leggeri who resigned in 2022. Until the appointment of the next Executive Director, Aija Kalnaja, previously working for the Latvian Police, has assumed the lead of Frontex being at the time the most senior Deputy Executive Director in the agency.

Frontex's funding consists of revenues mostly from the EU budget. Other possible sources of income stem from Schengen countries' regular or voluntary contributions, grants, and fees charged for services provided. The inclusion of this last source, namely fees charged, adds to the agency's autonomy as it constitutes a revenue controlled solely by the agency and not by the EU budgetary planning and fund allocation.

On Frontex's independence, agencies are supposed to act free of all political influence (Busuioc, 2009: 600) so as to fulfil their technocratic role unhindered. So, they are both cursed and blessed, operating 'at arm's length from national and EU authorities' (Brenninkmeijer, 2021). Indeed, Frontex's work is not linked to the EU institutions (Guild & Bigo, 2010: 268).

Accordingly, the European Commission does not participate in the drafting of Frontex's operational plans or the conduct of Frontex's operational activities, such as its joint operations. Despite participating and voting in management board meetings, with two representatives, on many occasions the European Commission has been outvoted (Font, 2018: 279). This manifests the Commission's decreasing influence and Frontex's high autonomy, especially taking into account that voting is a regular decision-making practice within Frontex's management board meetings shaping, in turn, the agency's internal power balances (Font, 2018: 279–280).

The European Parliament's role within Frontex is even more limited. Although Frontex is accountable to it, reports to its Committees, and the Parliament exercises budgetary control over the agency, it is up to the management board to invite, if it wishes, an expert of the European Parliament to attend a management board meeting, but without voting rights. In fact, the European Parliament's representatives frequently criticise the lack of parliamentary oversight over Frontex as well as report frail human rights' guardianship. Yet, many suggestions made by the European Parliament for institutional amendments of Frontex have been rejected (Léonard, 2012: 159–162). Even European parliamentarians calling two years in a row for

Frontex's Executive Director to quit did not produce any result. Neither did their vote in 2021 to freeze part of Frontex's budget for the next year. Rather, Leggeri was forced to resign when he lost the support of Frontex's management board over the result of an investigation conducted by OLAF regarding the agency's role in human rights abuses and misconduct. Or maybe he chose to resign to ward off any disciplinary actions to be taken against him over these findings. All these are manifestations of the European Parliament's frail reign over Frontex.

The Council participates in Frontex by virtue of its intergovernmental nature. Member states constitute the agency's main stakeholders, as they control its workings through their participation on the management board. However, the government is not actually represented there. Actually, the government's position is not expressed at Frontex's management board via a Minister or a diplomat. Instead, Frontex's management board meetings involve the heads or senior staff of border authorities, such as the Police of the Border Guard Service. This is due to the agency's technocratic and specialised agenda, which hinders any representation attempt, and therefore control, from a different pool, such as the Ministries of the Interior.

For Colimberti (2008: 35–38), the management board's composition manifests member states' hold on the agency, as national experts' preferences align with those of their national ministries, including the political leadership institutionalised through the Minister appointed by the elected government. This description however does not account for the inner dynamics of representation. Contrary to Colimberti's point, the management board's composition, rather than enhancing the grip of national capitals, actually lessens member states' control over Frontex. Staff from border authorities do not have the same degree of grounded national interests compared to diplomatic staff, whose job is simply to convey and safeguard their government's position. Staff from border authorities execute their profession, that is, border control, based on their professional knowledge and expertise (Adler & Haas, 1992; Haas, 1992; Radaelli, 1997). Thus, their views and votes on the management board meetings express their border control experience and not solely their government's interests. After all, the person that participates in Frontex's meetings is not always selected by the Minister but a decision based on bureaucratic hierarchy. This means that Ministers or even Prime Ministers may not exercise such tight control over Frontex through their country's management board representation. In parallel, although it is up to the Council to approve or dismiss any Frontex mandate amendment through the adoption of Council Regulation, there is still room for Frontex to act on its own. For instance, being a bureaucratic and extremely specialised institution, it has developed its own organisational character and working methods, without the Council's seal. In this line, Frontex has the right to 'communicate on matters falling within the scope of its tasks on its own initiative' (Regulation, 2019). Moving from being the subject of member states' control, Frontex has become the actual controller, as it monitors and periodically assesses member states' ability to face border control challenges.

All these reflect that, irrespective of member states' power and their ex-ante and ex-post control mechanisms put in place during its formation (Colimberti, 2008), Frontex still savours a level of autonomy, which was not originally stipulated by its creators. This autonomy increases as the agency continues to expand both in quantity, namely in terms of staff and budget, and quality, gaining more powers.

Frontex's expansion

Since becoming operational in May 2005 during its inaugural management board meeting, which took place at Warsaw's Marriott Hotel, Frontex has not ceased to expand. From its first Frontex-led joint operation entitled 'Illegal Labourers', at the EU eastern border in December 2005, to the conduct of 16 joint operations in 2019, not only at the EU land, sea, and air borders but even outside the EU territory. From less than 50 members of staff in 2005 to 1,520 persons today (see Chart 2.1).[3]

From a budget of around €6 million in 2005 to more than €750 million in 2022 (see Chart 2.2).

From occupying only 4 floors in the Rondo ONZ skyscraper during its first years to its relocation to a whole skyscraper in Plac Europejski, whilst there are also plans for the construction of its own building. Moreover, apart from its Warsaw premises, it maintains staff and offices in Catania (Italy), Piraeus (Greece), and Brussels (Belgium).

Beyond the numerical as well as spatial expansion, Frontex has also grown organisationally, as shown in Frontex organigrams drawn from 2008

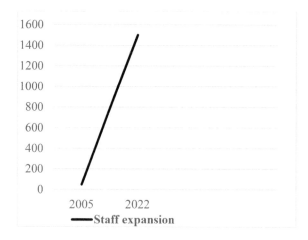

Chart 2.1 Frontex's staff expansion.

Frontex 29

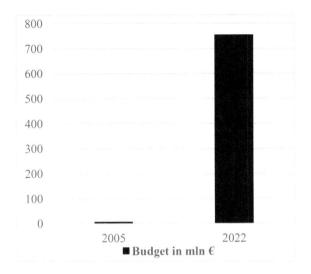

Chart 2.2 Frontex's budget expansion.

and 2021. Initially, maintaining a light structure (Colimberti, 2008: 29), it was composed of just 3 divisions and a few units (see Figure 2.1).

Today, marking a 200% increase, Frontex has 9 divisions equipped with various new units (see Figure 2.2).

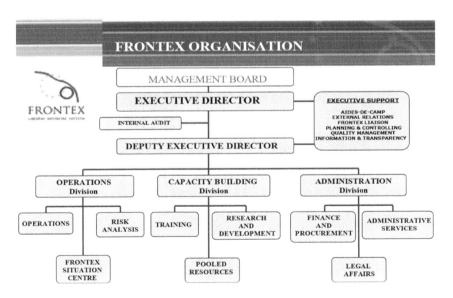

Figure 2.1 Frontex's organigram in 2008.
Source: ©Frontex.

30 *Frontex*

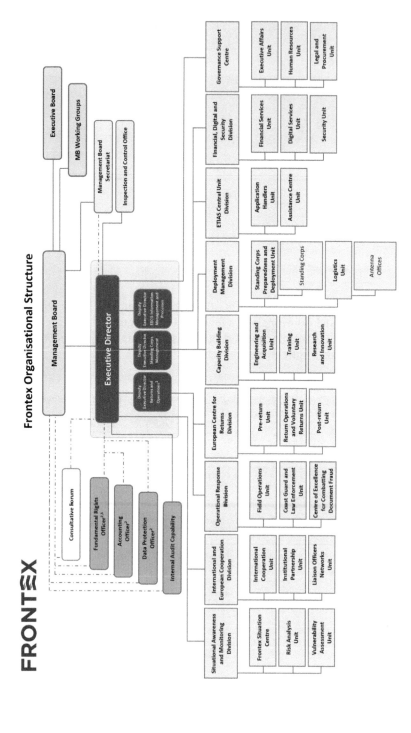

Figure 2.2 Frontex's organigram in 2022.
Source: ©Frontex.

This remarkable expansion matches with Frontex's mandate enhancement. Since its inaugural Regulation in 2004, Frontex has experienced 4 substantial mandate enhancements (2007, 2011, 2016, 2019). These enhancements brought new tasks, tools, and powers for this EU agency, which were not included in its founding act. For instance, the 2007 revision formed the Rapid Border Intervention Teams (RABIT), namely an emergency response mechanism for short-term deployment to a member state 'facing a situation of urgent and exceptional pressure' at its external borders due to a vast number of irregular entries (Regulation, 2007). The 2011 enhancement gave Frontex the right to purchase or rent its own technical equipment, increasing, in turn, the agency's autonomy, given that it is now able to acquire its own funds. In 2016, trying to accommodate its new responsibilities under the 2015 European Agenda on Migration (European Commission, 2015), Frontex was transformed into the European Border and Coast Guard (EBCG) agency, whereas the 2019 amendment granted Frontex a standing corps of 10,000 operational staff with executive powers and ready to intervene 'wherever and whenever needed' (European Commission, 2019). Analogously, Frontex's task increase is quite revealing. The agency's founding Regulation lists only 6 tasks (see Box B2.1). By contrast, its last mandate includes 33 tasks, manifesting a tremendous amplification of its scope of action (Regulation, 2019). Some of the new tasks assigned to Frontex include the provision of assistance at all stages of the return process, along with the monitoring on returns, cooperation with FRA for effective fundamental rights application, establishment of a communication network for secure and smooth exchange of sensitive data, operation of a False and Authentic Documents Online (FADO) system facilitating the detection of false travel documents as well as establishment and operation of the central unit of travel authorisation for

---

Box B2.1

Frontex's tasks as listed in its founding mandate (Regulation, 2004)

1 coordinate operational cooperation between member states in the field of management of external borders;
2 assist member states on training of national border guards, including the establishment of common training standards;
3 carry out risk analyses;
4 follow up on the development of research relevant for the control and surveillance of external borders;
5 assist member states in circumstances requiring increased technical and operational assistance at external borders;
6 provide member states with the necessary support in organising joint return operations

visa-exempt visitors travelling to Schengen (European Travel Information and Authorisation System — ETIAS).

The above indicate that Frontex is not the EU agency that was originally designed by its creators, namely the EU member states and the European Commission. It has expanded significantly both in measure and function. It has acquired new tools, more powers, new responsibilities, new staff, and increased financial resources. All these simultaneously strengthen and widen it as an institution and actor, aggravating, in turn, its paradoxical development.

Frontex's institutional brigade

On personnel, the agency is being staffed by 650 persons at the 'administrator' level and an equal number of persons at the 'assistant' level. Moreover, 220 officers have been seconded from their national authorities, whereas more than 900 persons work as contract agents, as attested by data included in Frontex's 2022 budget. Most staff are from Poland, followed by Romanians, Greeks, and Italians. On gender representation, approximately 70% are male officers, as shown in data drawn from Frontex's 2022–2024 programming document.

Besides the management structure and the Consultative Forum on Fundamental Rights, institutionally, Frontex's daily work is being allocated among its 9 divisions and 25 units (see Figure 2.2). Accordingly, the Situational Awareness and Monitoring division monitors and assesses the situation on the European borders through the work of the Frontex Situation Centre, the Risk Analysis Unit, and the Vulnerability Assessment Unit. Starting with the Frontex Situation Centre, its main task is to provide situational awareness with a real-time and constantly updated picture of EU borders. Although it has a different function and nature, as it deals with non-traditional security threats, its establishment follows the paradigm of pre-existing situation centres, such as of the North Atlantic Treaty Organisation (NATO), the Peacekeeping Situation Centre of the United Nations, and the EU Joint Situation Centre (Léonard, 2010: 243). Yet, those Centres mainly alert on developments posing a threat to peace. Now, Risk Analysis Unit identifies risks and key trends via data analysis so as to provide effective operational response to emerging challenges. Risk analysis is divided into Strategic Analysis, Operational Analysis, and Third Country Analysis. Vulnerability Assessment constitutes a recent unit set up to carry out vulnerability assessments on member states' capacity to manage their borders. These assessments can have the form of annual baseline assessments or specific assessments, which take into account the type and level of threats to which each member state is exposed and their possible impact. Besides assessing vulnerabilities, the unit recommends measures that would limit vulnerabilities, adopting a future-oriented approach.

The International and European Cooperation Division is composed of the International Cooperation Unit, the Institutional Partnership Unit, and the Liaison Officers Networks Unit. In sum, the division builds and enhances Frontex's relations with external partners, such as third countries like Canada, or international organisations, for example, the International Organisation for Migration or EU agencies like the European Union Agency for Law Enforcement Training (CEPOL). In doing so, it deploys liaison officers in third countries like Turkey, Ukraine, and Niger as well as EU member states, for instance, Greece, Bulgaria, France, and Belgium.

The Operational Response Division is the most visible Frontex division. It includes the Field Operations Unit, which plans, conducts, monitors, and evaluates Frontex's joint operations and rapid interventions at the borders. Moreover, there is a Coast Guard and Law Enforcement Unit, which focuses on the implementation of law enforcement and coast guard functions, and a Centre of Excellence for Combatting Document Fraud responsible for developing tools and services for document and identity control.

The European Centre for Returns is composed of a Pre-Return Unit involving, for instance, the implementation of EU readmission agreements, a Return Operations and Voluntary Returns Unit, which deals with the conduct of return operations, and a Post-Return Unit that engages with reintegration of returnees. Previously, Return Support constituted part of the Operations Division. Yet, it became a whole new division because of the need to build a stronger role for Frontex in the field of returns, as envisioned in its 2019 mandate revision.

The Capacity Building Division is divided into the Engineering and Acquisition Unit, the Training Unit, and the Research and Innovation Unit. The Engineering and Acquisition Unit, formerly called Pooled Resources Unit, is responsible for equipment procurement, that is, the acquisition and management of technical equipment and services for Frontex's operational needs. The Research and Innovation Unit brings innovation by creating, testing, evaluating, proposing, and buying new products for border control. In that regard, the unit advises member states on new technology as well as communicates with companies, organising industry meetings and functioning as an industry contact point. The Training Unit develops as well as harmonises education and training for border guards. One of its core tasks included the development of a Common Core Curriculum for border guards.

The Deployment Management Division is another new division created to manage the agency's standing corps, including its operational preparedness, deployment, composition with specific profiles, such as debriefing officers or forced return escorts, and the provision of logistical support. Frontex's standing corps is composed of border guards directly employed by Frontex, officers seconded from member states either for a short- or long period, and a rapid reserve pool with officers placed at the immediate disposal of Frontex in case of an emergency. Currently, Frontex's standing corps comprises 600 members, whereas it is planned to reach 10,000 by 2027.

The ETIAS Central Unit Division was established after the decision to grant Frontex the responsibility to set up and run the ETIAS Central Unit. The European Travel Information and Authorisation System (ETIAS) follows the path of the US Electronic System for Travel Authorisation (ESTA) and is due to become operational in the coming months. The division's task is to establish, develop, and manage the ETIAS Central Unit, which would operate on a 24/7 basis.

Lastly, there are two administrative divisions: the Financial, Digital, and Security Division and the Governance Support Centre, both of which support Frontex's administrative work, focusing on Frontex employees' needs.

Frontex's deeds

Frontex's mission is to 'ensure safe and well-functioning external borders providing security' (Frontex, 2022). Thus, it aims at protecting the EU citizens from any cross-border threat or crime, like irregular border crossings, overstays, false or falsified documents, stolen vehicles, smuggling of goods, weapons, and people or terrorism. To do so, it proposes, plans, coordinates, and evaluates joint operations at the EU air, land, and sea external borders or outside the EU territory. In case of a crisis situation, it deploys RABIT (also known as RBI) for immediate but short-term border interventions. Furthermore, it conducts return operations of illegally staying non-EU nationals. To conduct its joint operations, Frontex has created a pool of available experts and equipment ready to be deployed depending on the profiles listed in the operational plan. During operational activities, it oversees member states' compliance with fundamental rights. In parallel, it monitors the situation at the EU borders on a 24/7 basis, collecting data, analysing risks, and assessing vulnerabilities in order to act promptly in the event of an emergency or repair EU border control weaknesses. At hotspot areas, it provides operational assistance with screening, debriefing, and identification.[4] It trains national border guards and develops common training standards. It creates electronic platforms for information-sharing and border-related incidents' reporting. It runs electronic systems, such as the European Border Surveillance system (EUROSUR) and ETIAS. It collects and processes personal data. It implements and promotes the European Integrated Border Management as well as develops new strategies for border control. It publishes reports, even on ethics, develops glossaries, strategic analyses, and conducts research. It identifies priorities as well as proposes innovative tools and solutions for border management. It acquires its own border control assets. It cooperates with the private sector and the research industry. It organises events and liaises with border control authorities of member states and third countries.

The above constituted a brief presentation of Frontex's work. By way of illustration, some of its main tasks and functions are displayed schematically in Figure 2.3. Through this schema, it is possible to discern the diversity and broad range of Frontex's tasks. In parallel, these tasks indicate a multidimensional yet central function, which transcends the narrow 'EU agency' title conferred by its creators and defined in its legal mandate.

border control	migration	security	risk analysis
European Integrated Border Management	border surveillance	vulnerability assesments (& assessments)	EUROSUR
joint operations (air, sea, land borders)	rapid border interventions & return interventions	fundamental rights	monitoring
search & rescue	deployment of teams - standing corps	pool for technical equipment	training & education
support at hotspots	own capabilities & building capabilities	crossborder crime	terrorism
document fraud	returns	data collection	coast guard functions
ETIAS	research & innovation	development of technical standards	information exchange
cooperation with 3rd countries (incl. conduct of operations)	cooperation with national authorities	cooperation with EU partners	networks
liaison officers	secondary movements	best practices / guidelines / ethics	publications

Figure 2.3 Frontex's main tasks.

Indeed, institutionally, Frontex is an EU agency. This, however, does not indicate an abolition of actorness. EU agencies are actors (Groenleer & Gabbi, 2013; Rozée et al., 2013; Coman-Kund, 2018). Similarly, border control actors are not exclusively national states. After all, border control does not remain a state monopoly (Walters, 2006; Rumford, 2009). Following that, apart from being a border control EU agency, Frontex has also evolved into an actor implementing and shaping EU border control, that is, a border control actor. It is not just present at the EU borders and does not solely implement EU policies. Instead, it 'rules' over the EU border control regime (Vollmer & von Boemcken, 2014: 61), bringing new knowledge, attitudes, and values. Conducting joint missions and operational tasks, it formulates and executes operational decisions. It produces operational plans and coordinates joint missions taking tactical decisions about the deployment of resources during the conduct of missions. Furthermore, all Frontex staff and officers participating in Frontex operations are bound by Frontex's Code of Conduct. Deployed officers wear an armband bearing the insignia of the EU and Frontex, whereas members of its standing corps wear Frontex's uniform. In fact, Frontex has become the first EU service with its own uniform. Moreover, Frontex produces knowledge through its various activities, such as the organisation of workshops and events, risk analysis publications, data dissemination, and launch of operational measures. In parallel, Frontex is the sole EU-level actor on the field, namely at the EU borders, and, therefore, the EU institutions or even national authorities rely upon Frontex to stay informed. Similarly, they rely on Frontex's assessments, on Frontex's data, and operational actions. Thus, Frontex is not any more a 'tool' or the 'most important tool' as characterised by the former EU Commissioner for Migration and Home Affairs, Dimitris Avramopoulos. Rather, from a 'tool' for EU border control, Frontex has evolved into its key actor.

Frontex's role in EU border control: reviewing the literature

Despite the abundance of literature on Frontex, few accounts explore Frontex's impact on EU border control. These can be categorised into three main streams: the security stream, the institutionalist stream, and the border stream.

The security stream

Seeking to grasp Frontex's function in EU border control, various scholars focus on Frontex's link to security, viewing Frontex as a securitisation tool. Accordingly, Léonard (2010) and later Léonard and Kaunert (2022) consider Frontex's activities as securitising practices, which contribute to the securitisation of EU border control and migration. In particular, Léonard (2010), adopting a sociological perspective to securitisation advanced by Bigo (2002), analyses securitisation processes focusing on practices. She concludes that the main Frontex's activities can be considered to be securitising practices contributing therefore to the ongoing securitisation of asylum and migration in the EU (Léonard, 2010). 12 years later, Léonard and Kaunert (2022) reengage with practices in order to assess the security practices of Frontex during the two migration crises in the Mediterranean. They argue that there has been a qualitative change in Frontex's practices manifesting an intensification, or put differently, a spiralling of the securitisation of migration in the EU. Similar conclusions are drawn by Chillaud (2012) and Wiermans (2012), who, by examining EU discourses and practices, argue that Frontex supports the EU process of migration securitisation. This security-migration nexus in EU policy is also underscored by Horii (2016). Drawing on Bigo's concept of risk (2002), Horii looks into Frontex's risk analysis task. Exploring how Frontex's risk analysis is used in EU decision-making process, she finds that it is used as a tool, which adds to the legitimacy of European Commission's decisions.

The institutionalist stream

Other accounts, rooted in the discipline of EU governance, examine Frontex as a policy instrument so as to assess the institutionalisation of EU border control policy. In this context, Horii (2012) traces Frontex's training activities, whereas Paul (2017) explores Frontex's risk analysis function. In particular, Horii (2012), inspired by the ideas of sociological institutionalism and drawing on institutional isomorphism, elaborates on the effect of Frontex on the EU border management field, drawing attention to intra-organisational behaviour and institutional homogenisation. Paul (2017) applies an interpretive policy analysis so as to examine risk analysis as an institutionalising governance tool that harmonises European border control. Both reach the same verdict, namely that Frontex advances EU integration. In opposition, for Wolff and Schout (2013) Frontex is 'more of the same'. Building upon Scharpf's (1999) understanding of legitimacy, Wolff and Schout (2013) assess Frontex as an agency structure of EU governance. They contend that Frontex has not been a major addition to the EU's border control, as it did not result in its legitimisation. From a legal prism, Fernández-Rojo (2021) studies the implementation of EU border management and migration policies by EU agencies, namely Frontex, the European Union Agency for Asylum, and the European Union Agency for Law Enforcement Cooperation. Drawing on their legal mandates and assessing their reinforced operational tasks, he notes a vague legal framework that defines agency expansion and inter-agency function in these three agencies.

The border stream

Frontex's role in border re-territorialisation preoccupies works situated in the critical paradigm. In this line, Vaughan-Williams (2008) studies the surveillance activities of Frontex employed during operation Hera in the Canary Islands and Africa, applying the notion of 'borderwork' (Rumford, 2009). He identifies Frontex's surveillance activities as forms of an EU 'borderwork', which erodes the internal/external distinction. According to Vaughan-Williams (2008), the surveillance strategies employed during operation Hera can be considered as emerging bordering practices, which produce a border outside the space of border. Similarly, Reid-Henry (2013), adopting a Foucauldian logic, traces processes of re-territorialisation in Frontex's sea operations Hera and Nautilus in North Africa, arguing that Frontex fosters processes of extra-territorialisation and intra-territorialisation across the EU borders. For Reid-Henry (2013), Frontex is putting into place an 'incorporating geopolitics' of the border, which creates a context of politico-legal exception across borders. According to Reid-Henry (2013), through policy experimentation, Frontex incorporates in the EU's border control regime practices of neighbouring countries that can transform the EU's values and conception of democracy. In this context, goals are considered as more important than ethics, ideals, and normative values. Equally, Perkins and Rumford (2013) point out bordering processes during Frontex's joint operations in Africa and the Mediterranean, which attest borders' cosmopolitanisation. Approaching Frontex under a vernacularised border studies prism, they highlight the agency's practices of selective fixing and unfixing of border components during its joint operations, which shift the space of bordering activities away from the actual geographic location of borders. At the same time, echoing Reid-Henri (2013), they note that Frontex opens up the border to importing non-EU influences and practices, overlooking human rights breaches from third countries.

Towards Frontex's reinvigoration: bringing culture in

The above accounts, though they examine Frontex's role in EU border control, do not focus on Frontex. They concentrate instead on EU migration securitisation (Léonard, 2010; Chillaud, 2012; Wiermans, 2012; Horii, 2016; Léonard & Kaunert, 2022), EU border control policy institutionalisation (Horii, 2012; Wolff & Schout, 2013; Paul, 2017; Fernández-Rojo, 2021), or border re-territorialisation (Vaughan-Williams, 2008; Reid-Henry, 2013; Perkins & Rumford, 2013). In these analyses, Frontex is just a border control instrument or vehicle (Reid-Henry, 2013: 200), which exemplifies EU or national border control policies, downplaying thereby Frontex's role in EU border control. In other words, Frontex is considered as just a manifestation of border control and not an actor that can cause variation in the conduct of border control. This book seeks to remedy this, by zooming into Frontex as an actor capable of its own action and reaction. In other words, this book considers Frontex as an actor that can impact the EU border control.

To search for this impact, I turn to culture. As was shown above, so far culture has not been included in the scholarly work on Frontex. This constitutes an important omission, as culture interlinks the material world with the non-material, ideas with practices, conceptual conviction with empirical reality, theory with praxis, agency with structure, and even future with past. So, the insertion of culture opens up the road for the development of a different understanding of Frontex, which takes into account the social world and its agents. This, in turn, allows studying Frontex's interaction with the other actors of border control. After all, cultural predispositions shape social interaction and action (Bourdieu, 1993).

Accordingly, Frontex is not solely an institutional entity that operates at the EU external border and deals with border or security issues. It has a multifaceted function, which exceeds its legal mandate or its institutionally defined border control tasks. As an actor and agent, it operates in a social context producing social relations. This means that Frontex is a social actor embedded in a social environment. This constitutes another hidden aspect of Frontex's function, which can be explored with the invocation of culture. Accordingly, frequent interaction, common social activities, and the deepening of social relations among different actors can lead to the development of a community, whose members share certain collective cultural traits and generate a collective intentionality (Searle, 1995).

Actually, Frontex, aside from being a border control actor, is also a member of the border control community. This twofold function denotes that Frontex's role in EU border control transcends its technocratic mandate set by its creators. Frontex, as an actor of this social environment, is shaping ideas, perceptions, and beliefs, while, at the same time, is producing knowledge through its various activities, such as the suggestion and launch of operational measures, organisation of workshops and events, data dissemination, and risk analysis publications. Simultaneously, it interacts, communicates, and shares its experiences with the other members of the border control community developing, in turn, new understandings and operational activities. All these mean that Frontex also affects the border control community through its practices and ideas. This signifies a different role for Frontex in EU border control, that is, its participation in the border control community and interaction with the other members. In addition, this constitutes another literature lacuna, as the existence of a border control community has not been included in the relevant literature.

Hence, the inclusion of culture allows unveiling hidden aspects of Frontex's function in EU border control by opening up Frontex's research to new elements, such as cultural influences, ideational processes, and social contexts. Adopting this prism, Frontex moves away from its classification as an EU agency that blindly follows EU or national policies. Instead, Frontex becomes analysed as an actor that has a life on its own. To be more exact, an insurgent border control actor that rules over the EU borders by pursuing and promoting a new border control culture. Hence, Frontex's insurgency does not solely refer to its significant expansion beyond its creators' initial mandate and will or its autonomy and capacity to exercise control over member states. Frontex's insurgency also involves its pursuit of a new border control culture that has contested the Schengen regime, as the next chapters of this book will show.

Conclusion

This chapter helped us to get to know Frontex, the EU border control agency that since 2005 has become a permanent fixture at the EU external borders, whereas currently it has even expanded its operational activities outside of the EU soil. Characterising it as both a unique case in the EU institutional mechanism and an insurgent border control actor, the chapter argued that Frontex is not just another EU agency. Looking into the reasons that led to its birth, its governance, and institutional structure, the tasks entrusted upon it, and its considerable and non-stop expansion, it tried to provide a holistic overview of this controversial agency. Then, presenting and reviewing the scholarly literature that engages with Frontex's impact on European border control, it identified certain important gaps, which this book seeks to address with the insertion of culture to Frontex's analysis. In doing so, it applies a cultural approach aiming at scrutinising Frontex's role as a border control actor embedded in a border control policy community at the EU external borders. The next chapter inserts culture into Frontex's analysis via the application of the 'cultures of border control' analytical framework (Zaiotti, 2011).

Notes

1 Deciding to materialise the aspiration for a common market set with the Treaty of Rome in 1957 and expand free mobility of workers to free movement of persons, the leaders of the then 12-member states of the European Communities, endorsed in 1986 the Single European Act. The adoption of this document promoted European integration and constituted a decisive step towards the harmonisation of national policies, as it established an area without internal frontiers with the removal of all barriers to the free flow of persons, goods, services, and capital. For the negotiations and reform of the Single European Act, see Moravcsik (1991).
2 Ireland can attend management board meetings but without voting rights, as it does not participate in the Schengen Agreement. Limited voting rights have the Schengen Associated Countries, namely Norway, Iceland, Lichtenstein, and Switzerland.
3 See next part for staff breakdown.
4 The hotspot approach is part of the EU immediate action to assist operationally certain frontline EU member states that face disproportionate migratory pressures (European Commission, 2015).

References

Adler, E. & Haas, P.M. (1992) 'Conclusion: Epistemic Communities, World Order, and the Creation of a Reflective Research Program'. *International Organization*, 46(1), pp. 367–390.

Bigo, D. (2002) 'Security and Immigration: Toward a Critique of the Governmentality of Unease'. *Alternatives*, 27(1), pp. 63–92.
Bourdieu, P. (1993) *The Field of Cultural Production*. Cambridge: Polity Press.
Brenninkmeijer, A. (2021) 'The Unknown Agents of European Cooperation, and their Future'. *ERA Forum*, 22, pp. 245–251.
Busuioc, M. (2009) 'Accountability, Control and Independence: The Case of European Agencies'. *European Law Journal*, 15(5), pp. 599–615.
Chillaud, M. (2012) 'Frontex as the Institutional Reification of the Link between Security: Migration and Border Management'. *Contemporary European Studies*, 2, pp. 45–61.
Colimberti, F. (2008) *Frontex: A Principal-Agent Perspective*. Munich: GRIN.
Coman-Kund, F. (2018) *European Union Agencies as Global Actors: A Legal Study of the European Aviation Safety Agency, Frontex and Europol*. Abingdon: Routledge.
Commission of the European Communities (2003a) 'Proposal for a Council Regulation establishing a European Agency for the Management of Operational Co-Operation at the External Borders'. COM(2003)687.
Commission of the European Communities (2003b) 'Presidency Conclusions: Thessaloniki European Council'. Available at https://ec.europa.eu/commission/presscorner/detail/en/DOC_03_3 (accessed January 2022).
Diez, T., Stetter, S. & Albert, M. (2006) 'The European Union and Border Conflicts: The Transformative Power of Integration'. *International Organization*, 60(3), pp. 563–593.
Ekelund, H. (2014) 'The Establishment of FRONTEX: A New Institutionalist Approach'. *Journal of European Integration*, 36(2), pp. 99–116.
European Commission (2015) 'A European Agenda on Migration'. COM(2015)240.
European Commission (2019) 'European Border and Coast Guard: The Commission Welcomes Agreement on a Standing Corps of 10,000 Border Guards by 2027'. Available at https://ec.europa.eu/commission/presscorner/detail/en/IP_19_1929 (accessed January 2022).
European Court of Auditors (2021) 'Frontex's Support to External Border Management: Not Sufficiently Effective to Date'. Special Report 08.
Fernández-Rojo, D. (2021) *EU Migration Agencies: The Operation and Cooperation of FRONTEX, EASO and EUROPOL*. Cheltenham: Edward Elgar.
Font, N. (2018) 'Informal Rules and Institutional Balances on the Boards of EU Agencies'. *Administration & Society*, 50(2), pp. 269–294.
Frontex (2010) *Beyond the Frontiers*. Warsaw: Frontex.
Frontex (2022) 'Our Mission'. Available at https://frontex.europa.eu/about-frontex/our-mission/ (accessed February 2022).
Groenleer, M. (2009) *The Autonomy of European Union Agencies: A Comparative Study of Institutional Development*. Delft: Eburon.
Groenleer, M. & Gabbi, S. (2013) 'Regulatory Agencies of the European Union as International Actors: Legal Framework, Development over Time and Strategic Motives in the Case of the European Food Safety Authority'. *European Journal of Risk Regulation*, 4(4), pp. 479–492.
Guild, E. & Bigo, D. (2010) 'The Transformation of European Border Controls'. In: B. Ryan & V. Mitsilegas (eds.), *Extraterritorial Immigration Control: Legal Challenges*. Leiden: Martinus Nijhoff, pp. 257–279.
Haas, P.M. (1992) 'Introduction: Epistemic Communities and International Policy Coordination'. *International Organization*, 46(1), pp. 1–35.

Horii, S. (2012) 'It Is About More than Just Training: The Effect of Frontex Border Guard Training'. *Refugee Survey Quarterly*, 31(4), pp. 158–177.

Horii, S. (2016) 'The Effect of Frontex's Risk Analysis on the European Border Controls'. *European Politics and Society*, 17(2), pp. 242–258.

House of Lords (2003) 'Proposals for a European Border Guard'. *Select Committee on the European Union*, HL Paper 133.

Jorry, H. (2007) 'Construction of a European Institutional Model for Managing Operational Cooperation at the EU's External Borders: Is the FRONTEX Agency a Decisive Step Forward?'. *CEPS*, Challenge Research Paper 6, pp. 1–32.

Kasparek, B. (2010) 'Borders and Populations in Flux: Frontex's Place in the European Union's Migration Management'. In: M. Geiger & A. Pécoud (eds.), *The Politics of International Migration Management*. Basingstoke: Palgrave Macmillan, pp. 119–140.

Léonard, S. (2009) 'The Creation of FRONTEX and the Politics of Institutionalisation in the EU External Borders Policy'. *Journal of Contemporary European Research*, 5(3), pp. 371–388.

Léonard, S. (2010) 'EU Border Security and Migration into the European Union: Frontex and Securitisation through Practices'. *European Security*, 19(2), pp. 231–254.

Léonard, S. (2012) 'The Role of Frontex in European Homeland Security'. In: C. Kaunert, S. Léonard & P. Pawlak (eds.), *European Homeland Security: A European Strategy in the Making?* Abingdon: Routledge, pp. 145–164.

Léonard, S. & Kaunert, C. (2022) 'The Securitisation of Migration in the European Union: Frontex and Its Evolving Security Practices'. *Journal of Ethnic and Migration Studies*, 48(6), pp. 1417–1429.

Majone, G. (2006) 'Managing Europeanization: The European Agencies'. In: J. Peterson & M. Shackleton (eds.), *The Institutions of the European Union*. Oxford: Oxford University Press, pp. 190–209.

Marenin, O. (2010) 'Challenges for Integrated Border Management in the European Union'. *DCAF*, Occasional Paper 17, pp. 1–161.

Monar, J. (2006) 'Cooperation in the Justice and Home Affairs Domain: Characteristics, Constraints and Progress'. *European Integration*, 28(5), pp. 495–509.

Moravcsik, A. (1991) 'Negotiating the Single European Act: National Interests and Conventional Statecraft in the European Community'. *International Organization*, 45(1), pp. 19–56.

Neal, A. (2009) 'Securitization and Risk at the EU Border: The Origins of FRONTEX'. *Journal of Common Market Studies*, 47(2), pp. 333–356.

Paul, R. (2017) 'Harmonisation by Risk Analysis? Frontex and the Risk-Based Governance of European Border Control'. *Journal of European Integration*, 39(6), pp. 689–706.

Perkins, C. & Rumford, C. (2013) 'The Politics of (Un)fixity and the Vernacularization of Borders'. *Global Society*, 27(3), pp. 267–282.

Pollak, J. & Slominski, P. (2009) 'Experimentalist but Not Accountable Governance? The Role of Frontex in Managing the EU's External Borders'. *West European Politics*, 32(5), pp. 904–924.

Radaelli, C.M. (1997) *The Politics of Corporate Taxation in the European Union: Knowledge and International Policy Agendas*. Abingdon: Routledge.

Regulation (EC) No 2007/2004 of 26 October 2004 Establishing a European Agency for the Management of Operational Cooperation at the External Borders of the Member States of the European Union [2004, OJ L 349/1].

Regulation (EC) No 863/2007 of the European Parliament and of the Council of 11 July 2007 Establishing a Mechanism for the Creation of Rapid Border Intervention Teams and Amending Council Regulation (EC) No 2007/2004 as regards that Mechanism and Regulating the Tasks and Powers of Guest Officers [2007, OJ L 199/30].

Regulation (EU) 2016/1624 of the European Parliament and of the Council of 14 September 2016 on the European Border and Coast Guard and Amending Regulation (EU) 2016/399 of the European Parliament and of the Council and Repealing Regulation (EC) No 863/2007 of the European Parliament and of the Council, Council Regulation (EC) No 2007/2004 and Council Decision 2005/267/EC [2016, OJ L 251/1].

Regulation (EU) 2019/1896 of the European Parliament and of the Council of 13 November 2019 on the European Border and Coast Guard and Repealing Regulations (EU) No 1052/2013 and (EU) 2016/1624 [2019, OJ L 295/1].

Regulation (EU) No 1168/2011 of the European Parliament and of the Council of 25 October 2011 Amending Council Regulation (EC) No 2007/2004 Establishing a European Agency for the Management of Operational Cooperation at the External Borders of the Member States of the European Union [2011, OJ L 304/1].

Reid-Henry, S.M. (2013) 'An Incorporating Geopolitics: Frontex and the Geopolitical Rationalities of the European Border'. *Geopolitics*, 18(1), pp. 198–224.

Rijpma, J. (2009) 'EU Border Management after the Lisbon Treaty'. *Croatian Yearbook of European Law and Policy*, 5, pp. 121–149.

Rozée, S., Kaunert, C. & Léonard, S. (2013) 'Is Europol a Comprehensive Policing Actor?'. *Perspectives on European Politics and Society*, 14(3), pp. 372–387.

Rumford, C. (ed.) (2009) *Citizens and Borderwork in Contemporary Europe*. Abingdon: Routledge.

Scharpf, F. (1999) *Governing in Europe: Effective and Democratic?*. Oxford: Oxford University Press.

Searle, J.R. (1995) *The Construction of Social Reality*. London: Penguin.

Vaughan-Williams, N. (2008) 'Borderwork beyond Inside/Outside? Frontex, the Citizen-Detective and the War on Terror'. *Space and Polity*, 12(1), pp. 63–79.

Vollmer, R. & von Boemcken, M. (2014) 'Europe off Limits: Militarization of the EU's External Borders'. In: I. Werkner, J. Kursawe, M. Johannsen, B. Schoch & M. von Boemcken (eds.), *Friedensgutachten 2014*. Münster: LIT, pp. 57–69.

Walters, W. (2006) 'Border/Control'. *European Journal of Social Theory*, 9(2), pp. 187–203.

Wiermans, M. (2012) *The Securitization of Frontex: A Discours Analysis*. Saarbrücken: Lambert.

Wolff, S. & Schout, A. (2013) 'Frontex as Agency: More of the Same?'. *Perspectives on European Politics and Society*, 14(3), pp. 305–324.

Zaiotti, R. (2011) *Cultures of Border Control: Schengen & the Evolution of European Frontiers*. Chicago: University of Chicago Press.

3 Constructing Frontex's culture

'Frontex identified a demand for strategic leadership that will support [...] a common EU border and coast guard culture'.

—Frontex (2017a)

Introduction

Agencies are not just technocratic structures created to fulfil their creators' wishes. They can evolve into value-bearing institutions, which develop their own ideas and opinions about what is appropriate behaviour (Lægreid et al., 2011: 1328) and how they are being or should be self-represented. Moreover, they can emancipate themselves to avoid unpleasant or dysfunctional situations (Copeland, 2000: 190). As such, apart from impacting the material world, whilst operating and performing the tasks assigned to them, they can also shape their social environment through social interaction, moving, in turn, the attention to social dynamics and social relations that mould and co-shape the non-material context of the social world. In this sense, agencies act and behave like social actors, which generate a collective intentionality (Searle, 1995: 23–26), by internalising, producing, and re-producing institutional rules, norms, and cultural values (Schimmelfening, 2003: 69). On that account, they have the ability to interpret and construct a new social reality.

Following this spirit, the chapter inserts culture into Frontex's analysis. Applying the 'cultures of border control' analytical framework (Zaiotti, 2011), it explores Frontex's cultural disposition. The argument advanced is that Frontex has developed its own cultural traits, which are composed of specific assumptions and practices for borders and border control. These assumptions and practices are part of its culture. So, Frontex is also an actor capable of developing and pursuing its own culture, shaping therefore its social environment. The research is based on document analysis, institutional discourse analysis, and interviews with Frontex staff in Warsaw.

Structurally, the chapter is divided into two sections. The first section refers to Frontex's role as a border control actor and central position in the border control community, which has been consolidated due to the

DOI: 10.4324/9781003230250-3

recognition that it has gained from the other border control actors, border control stakeholders, third countries as well as its operational results. The second section presents Frontex's border control assumptions and practices. The chapter argues that these assumptions and practices reflect the workings of Frontex and specifically denote the content and direction of Frontex's border control role.

Frontex: the essential EU border control actor

Since its operational inauguration in 2005, Frontex has accumulated a significant operational role, expertise, and border control knowledge (Pollak & Slominski, 2009: 908; Rijpma, 2012: 90; Horii, 2016; Campesi, 2022: 138–147). Actually, it does not just implement but, most importantly, it formulates border policy.

Each month, around 1,200 Frontex officers are deployed at EU borders (Frontex, 2022b). Indeed, the numbers are indicative of Frontex's operational involvement in EU border control. In 2019, Frontex conducted 16 joint operations at the external land, air, and sea borders (Frontex, 2020b: 14), reporting 139,000 illegal border crossings, detecting more than 6,000 falsified documents, rescuing more than 54,000 border crossers, returning more than 15,000 non-EU nationals (Frontex, 2020b: 4–5), and training more than 3,600 officers (Frontex, 2020b: 60). I chose to refer to 2019, that is, before the COVID-19 outbreak, which unavoidably put a dent in Frontex's operational planning by limiting the number of officers deployed at the borders.

Besides, Frontex has started acquiring its own equipment to be deployed during the conduct of its joint operations so as to limit its material dependency upon its member states, such as mobile surveillance systems, patrol cars, and vehicles for migration management support used as mobile offices. Next in Frontex's acquisition list are vessels, planes, and drones.

Its continuous presence at the borders with human and technical resources renders Frontex a border control actor: an actor that is highly visible whilst operating on the field. For instance, during Frontex land operations, border patrolling usually involves two Frontex officers and one national border officer from the host member state (Interviewee 7), attesting Frontex's numerical presence. Enhancing its visibility, Frontex's deployed resources bear Frontex's logo, such as on cars, or on the uniforms of the officers participating in its joint operations. Soon, that is, by 2027, Frontex will be ready to deploy 10,000 officers, forming its standing corps and dressed in Frontex's uniform.

Building on its border control role at the EU borders, it cooperates with national authorities and border or coast guards, which serve at their national borders. Interacting and sharing common experiences, participating in Frontex's joint operations and other operational actions, border and coast guards develop relations with their colleagues who serve as Frontex officers. So, they start developing a community, as the next chapters show.

Being present as an EU border control agency, Frontex functions as an initiator, shaper, filter, or even barrier to courses of action in the field of EU border management (Allen & Smith, 1990). In this sense, operationally, Frontex acts at the border by formulating and adopting operational decisions during the conduct of its joint operations. In particular, in cooperation with the host member state, it drafts operational plans, it determines missions' duration or missions' renewal, it allocates the budget, as well as selects the type or profile of deployed resources, including the acting staff equipped with executive powers (Regulation, 2019).

Moreover, it builds new forms of cooperation via the establishment of networks or partnerships, like the European Patrols Network, the Information and Coordination Network, the Frontex Risk Analysis Network (FRAN), and the Western Balkans Risk Analysis Network (WB-RAN). Creating these structures, Frontex gives life to new forms of cooperation and sets novel ways of acting, which constitute an actor's characteristic (Finnemore, 1996: 30). In parallel, being their creator, Frontex relishes a central role within these networks, fortifying, as a result, its position (Pollak & Slominski, 2009: 907).

Beyond that, it sets multiannual goals, strategic objectives, and a strategic direction (Frontex, 2021a: 14–17). It has articulated its vision, mission, and is promoting its own values, published in its institutional webpage (Frontex, 2022a), which reflect the agency's unique character as a bureaucratic agency and distinct 'personality' vis-à-vis other EU agents. It self-assesses its performance, sets priorities, and lists its key achievements (Frontex, 2021b). Apart from evaluating itself, it also evaluates EU member states' ability to safeguard their borders. In this context, Frontex monitors and evaluates member states' capacity and readiness to face challenges at their borders (Regulation, 2019). This means that the evaluee, namely the agency, became the evaluator of its creators.

As such, it is not just present at the EU borders and does not solely implement EU policies. Instead, it 'rules' over the EU border control regime (Vollmer & von Boemcken, 2014: 61), bringing new tools, knowledge, attitudes, and values. This alludes that Frontex, from an EU agency, has evolved into an actor. Its actorness is attested by the fact that Frontex is recognised and accepted as an interlocutor by other actors. Indeed, national border guards, states, international organisations, and agencies recognise, accept, and interact with Frontex. For instance, Frontex participates in EU-US Justice and Home Affairs Ministerial Meetings. It is present in the Organisation for Security and Co-Operation in Europe (OSCE) roundtables on border management as well as conducts visits for border management assessment in third countries (EEAS, 2016). During the EBCG Day of 2019, Frontex hosted approximately 600 border and coast guard officers from more than 30 countries (Frontex, 2019a). This number shows that Frontex has gained recognition and approval as a border control actor by its partners. This is also illustrated discursively:

[Interpol] 'Interpol and Frontex have a long history of cooperation'
(Frontex, 2017b).

[NATO] 'We decided to increase our cooperation with the EU and Frontex. And I'm very grateful that we have been able to really establish a very practical cooperation with Frontex'
(NATO, 2016).

[European Court of Auditors' Audit] 'The Agency has become an important actor in migration enforcement on the European stage'
(European Court of Auditors, 2020: 9).

In parallel, Frontex is being self-identified as a border control actor. It characterises itself as a 'key actor at the European level' (Frontex, 2019b) and 'a fully-fledged internal security actor' (Frontex, 2017c), underscoring its importance, as reflected in its Executive Director's rhetoric:

[Frontex Executive Director] 'Frontex has become an essential actor in law enforcement on the European level'
(Frontex, 2017c).

Frontex's actorness is being complemented by its key role, which leads it to be perceived and function as a 'primus inter pares' among other actors (Rijpma, 2016: 19), as shown below:

[EU Commissioner] 'The management of external borders has increasingly become a shared responsibility. Frontex, as a coordinator, will play an essential role in its implementation'
(Avramopoulos, 2015).

Moreover, it constitutes the primary source from which the EU institutions, and especially the European Commission and the European Parliament, derive statistical information and data about the situation at the EU external borders.[1] So, Frontex is not just a numerical supplier of information. Instead, it is deemed a reliable source upon which member states and EU institutions rely to get valid information and plan their future course of action. Indeed, the European Commission asks Frontex to draft reports and proposals on issues on which it plans to undertake actions (Interviewee 1).

[Frontex officer] 'We are doing it because they asked us. They did not have anyone else to do it'
(Interviewee 2).

It is also considered rather effective, taking into account that member states and even third countries continue to ask for Frontex's involvement when they face increased migratory pressures.[2] Even the UK, amidst Brexit, was not averse to Frontex's border control involvement in Gibraltar (EURACTIV, 2021). That means that it approves of Frontex's work and is not averse to its border control contribution. After all, well before exiting the EU, the UK, though it was not a signatory of the Schengen Agreement, had sought full membership in Frontex, distinguishing it in practice from Schengen (House of Lords, 2017).[3]

Hence, Frontex does not only participate in the border control conduct and it does not solely facilitate national border control policies. It has gained acceptance, recognition, and credibility among its partners. Progressively, this recognition and credibility can lead to causal effects (Zürn & Checkel, 2005: 1049) making this agency more persuasive towards other actors. Accordingly, both the European Commission and its member states rely upon Frontex's expertise to formulate border control or even migration management policies. They also use Frontex's data and analyses to justify their decisions. For instance, the European Commission justified budget allocation of the External Borders Fund based on data drawn from Frontex (Horii, 2016). They also rely upon Frontex to get informed about border vulnerabilities, potential risks, and shifts in border crossing routes (European Commission, 2020; Frontex, 2020c). They rely upon Frontex for the harmonisation of EU and national border control policies (Paul, 2017). They rely upon Frontex for the education and training of national and EU border guards or the deployment of innovative border control tools. At the same time, they accept and adopt Frontex's working methods, such as its strategic documents, codes of conduct, operational plans, and various publications for EU border control, including its ethics.[4] Furthermore, they approve and integrate in their national system information technology (IT) reporting systems and platforms developed by Frontex, such as the Joint Operations Reporting Application (JORA), the Frontex One-Stop-Shop (FOSS), and EUROSUR. Through these reporting platforms, which national authorities are obliged to use daily to upload data, Frontex promotes its own mode of operational reporting and visualisation. Thus, it cultivates its own border control logic. The methodology used to upload an incident, namely the title, words used, information included, and mode of presentation, reflects a specific operational logic that influences the conduct of border control. Hence, given that Frontex designed, implemented, and now leads these platforms, it preferentially advances its own mode of border control application.

The above show that Frontex is not just a border control agency but a border control actor, which produces both operational actions and intersubjective meanings. In parallel, engaging with border control and having a physical presence at the geographic location of the border, Frontex turns into a social actor. Its presence at the borders demands its interaction with the other border control actors, namely national border guards. This interaction renders Frontex an actor embedded in a community, namely a social actor which acts and interacts with the other border control actors. At the same time, this interaction manifests Frontex's acceptance by national border guards. Otherwise, Frontex would not have been able to establish new forms of cooperation, such as networks, or conduct joint operations staffed with national border guards or events that require border guards' participation, like the EBCG Day. Manifesting this, Frontex's call for just 700 border guards' posts received 7,500 applications (EUobserver, 2020). So, more than 10 candidates per post. The increased interest in Frontex's job vacancies shows that the agency projects enough incentives outside of the agency having the ability to attract old or aspiring border guards.

Frontex's participation in the border control conduct and its interaction with the other border control actors lead Frontex to function as a change actor or, put differently, a change agent (Börzel & Risse, 2003: 67–68) and a policy entrepreneur (Kingdon, 2014 [1984]). Aside from creating new border control tools, like platforms and risk analysis networks, as mentioned earlier, Frontex brings also its own ideas, practices, and activities, which it carries during its interaction with the other border control actors, transmitting them therefore in a dynamic process through socialisation. Socialisation is irrefutably a two-way process among the socialising agents, yet, the introduction of a new actor can shift established patterns of behaviour or taken for granted assumptions. Let alone if this actor functions as the glue among diverse border control actors or if it is considered the 'primus inter pares' among them. The actor, in this case Frontex, can initiate considerable change by persuading the other members of the community to redefine their interests or ideas and, in turn, transform border policy.

The above account for a different role for Frontex, acknowledging its behaviour, meanings, practices, and social environment in which Frontex is embedded. Indeed, they illustrate that Frontex is not a mere technocratic EU agency that just implements or follows EU policies. Rather, it is an actor that acts at its social environment, producing and reproducing behavioural patterns, meanings, and relations with the other border control actors and members of the border control community. Being involved with EU border control since 2005 and after 17 years of continuous operational action at the EU external borders, Frontex has become a mature border control actor. Actually, having reached maturity, Frontex has evolved into an actor on its own right, which pursues a 'modus vivendi and essendi'. As an actor, it can be a carrier of ideas and a producer of inter-subjective meanings (Saurugger, 2013: 898). Its behaviour can be shaped by socially shared understandings of the world (Berger, 2000: 410), such as identity, norms, and culture. Thus, the EU agency status does not inherently exclude Frontex from the development and/or pursuing of a culture. After passing its formative years and having reached maturity, Frontex can become an expressor and promoter of culture.

Delving into Frontex's cultural traits

Frontex, as a border control actor and border control practitioner, is composed of and, at the same time, produces assumptions and practices for the border control conduct. These assumptions and practices elucidate the inner dynamics and workings of the agency. Furthermore, they set the parameters of Frontex's action at the borders, informing therefore the border control conduct, due to Frontex's border control role. Background assumptions refer to border and border control characteristics, while practices are activities or actions. These assumptions and practices are systemically promoted by the agency and, as a result, are present in almost each of its activities. Otherwise, they would not have been constituting its cultural traits. Being part of its institutional culture there is no escape from them.

Starting with border assumptions, Frontex perceives borders as both a national territory and an EU space. The national aspect sets the parameters of Frontex's action on the ground. Accordingly, Frontex states that 'responsibility for border control [...] remains the exclusive competence of member states' (Frontex, 2012). Besides the national angle, there is also an EU context enabling Frontex to operate at its 'EU agency' capacity. Highlighting this dimension, Frontex often refers to borders as a single EU external border. The dual nature of borders prompts Frontex and national authorities to cooperate 'in a spirit of shared responsibility' (Regulation, 2019). After all, 'the protection of the area of the European area of freedom, security and justice is a shared responsibility of all member states and Frontex' (Frontex, 2020d).

Regarding border control assumptions, Frontex emphasises fundamental rights. This may sound paradoxical. After all, the most common criticism levelled against Frontex refers to human rights violations, as already mentioned in the book's opening pages. Yet, it seems that Frontex is actively and deliberately employing a discourse that emphasises fundamental rights and humanitarianism (Aas & Gundhus, 2015):

> [Frontex Executive Director] 'Fighting crime at the border is a key objective of the Lisbon Treaty, one of the cornerstones of which is full respect for fundamental rights'
>
> (Frontex, 2010b).

> [Frontex Executive Director] '[...] ensuring at all times that irregular migrants are properly identified and treated in line with our commitment to fundamental rights and human dignity'
>
> (Frontex, 2010c).

Frontex has articulated a Fundamental Rights Strategy, which sets its fundamental rights principles, stating in its preamble that the agency 'integrates fundamental rights safeguards throughout all its activities, at all stages' and that it is 'fully committed to developing, promoting, and ensuring [...] respect for fundamental rights' (Frontex, 2021c: 3). Hence, according to Frontex, the protection of fundamental rights guides its operational activities, being an unconditional component of effective border management (Frontex, 2021c). Organisationally, within Frontex operates a Fundamental Rights Officer, a Complaints Mechanism for the submission of fundamental rights violations, and a Fundamental Rights Consultative Forum. Furthermore, respect of fundamental rights constitutes a separate chapter in Frontex's Code of Conduct and a section in the Common Core Curriculum. As a result, all Frontex staff and deployed officers are informed and trained to act respecting fundamental rights. The promotion of fundamental rights indicates a human rights-based border control approach (Pascouau & Schumacher, 2014: 1). Yet, taking into account the number of incidents reported for potential fundamental rights violations during Frontex operations (Frontex, 2021d: 14–15), it is possible to argue that this human rights prioritisation reflects an aspiration or an organisational strategy of Frontex to reposition itself as a protector of humanitarianism (Perkowski, 2018) and not an implemented action always encountered and ensured during Frontex activities.

Ironically at odds with fundamental rights, Frontex's engagement with border control is based on a securitisation logic (Léonard, 2010; Chillaud, 2012; Wiermans, 2012; Horii, 2016; Léonard & Kaunert, 2022). The phrases of its Executive Director that 'Europe is only as secure as its external borders' (Frontex, 2018a) and that the agency 'remains committed and vigilant in its efforts to ensure security of Europe's borders' (Frontex, 2017d: 7) are just a few examples of Frontex's discursive endeavour to link border control with security. Similarly, in its webpage, Frontex articulates that its mission is to 'ensure safe and well-functioning external borders providing security' (Frontex, 2022a). Furthermore, it considers itself as a 'fully-fledged internal security actor' (Frontex, 2017c) and a promoter of 'a pan European model of integrated border security' (Frontex, 2007). Viewing security as integrated in its core 'business', in 2011, Frontex published a study on Ethics in Border Security, whereas, in 2021, Frontex, along with the League of Arab States, held a Euro-Arab Border Security Conference to discuss security concerns and enhance the operational cooperation on security issues. On securitisation, arguably, almost every Frontex activity is considered to be promoting the EU's securitisation (Léonard, 2010). A top example is the RABIT mechanism, designed to be activated in case of urgent and extraordinary circumstances. This securitisation emphasis is also manifested organisationally, for instance, through the function of the Risk Analysis Unit. This unit applies a Common Integrated Risk Analysis Model (CIRAM), which defines risk as a function of threat, vulnerability, and impact (Frontex, 2013a: 5). Assessing risks in terms of threats, this model can lead to the construction of new threats, functioning, therefore, as an enabler of securitisation (Buzan et al., 1998). Likewise, the Vulnerability Assessment Unit engages with the identification and assessment of possible security risks or threats at the EU borders to prevent crises. The unit's focus on preventive action against security crises indicates a latent securitisation. Moreover, turning to institutional terminology and document analysis, the agency in its webpage often uses the terms 'crime' and 'security threats' as well as the rather strong verbs of 'tackle', 'combat', and 'battle'. This terminology clearly manifests a securitisation logic as it is framed with a negative connotation underscoring threats, borders at risk, and security crises. Actually, in its 2021 Risk Analysis document, the word 'security' is mentioned 16 times and 'threat' 27 times. The spread of these terms illustrates that for Frontex borders are tied to security preoccupations diffusing as a result a securitisation logic.

Frontex also produces a re-territorialisation of the border. Perceiving that the place of the border control conduct is not geographically limited to the actual borderline, it shifts the border, redefining therefore the internal/external border dichotomy (Vaughan-Williams, 2008; Perkins & Rumford, 2013; Reid-Henry, 2013). That is reflected in the following institutional rhetoric:

> [Frontex Deputy Executive Director] 'Knowledge and control of what happens before the border in neighbouring countries and what happens inland, once the border has been crossed, is also of vital importance'
> (Frontex, 2010a: 79).

> [Frontex officer] 'We cannot win the battle of irregular migration at the borderline'
> (Interviewee 3).

> [Head of the Frontex Situation Centre] '[...] we need to go beyond the borders, we need to go to the pre-frontier area'
> (Frontex, 2019c).

This re-territorialisation is cotemporally projected via extra-territorialisation and intra-territorialisation. On the one hand, Frontex contributes to border control's extra-territorialisation. Its surveillance activities in the pre-frontier area, namely beyond the external border, extend border control to neighbouring countries' territory. The same applies with the deployment of liaison officers to third countries as well as the conduct of joint border patrolling with officers from neighbouring or even non-neighbouring states, like Mauritania (Frontex, 2010a: 32). Another extra-territorialisation example is ETIAS, which refers to a pre-travel authorisation. Non-EU citizens travelling to a Schengen country from a visa-free country need to apply for an ETIAS authorisation. These applicants will not be at the border to be checked while submitting their application but in their home country. So, operating this system, Frontex will filter potential travellers before even starting their journey. Yet, extra-territorialisation's most obvious paradigm constitutes Frontex's operations in non-EU countries, namely in Albania, Serbia, and Montenegro, that is, outside of Frontex's geographic area of responsibility. All these manifest that for Frontex border control is not spatially confined to the EU territory (Ferrer-Gallardo & van Houtum, 2014:

299). The same conclusion is drawn from Frontex's institutional terminology and its choice to frequently use in its publications, instead of 'border', the wording 'borderlands' and 'beyond the frontiers', which indicate a hybrid space, namely a zone extended away from the location of the border.[5] On the other hand, Frontex engages in border control activities within the EU territory, yet away from the actual border, denoting an 'intra-territorialisation' (Reid-Henry, 2013: 218). For instance, its return operations do not concern irregular border crossers but illegally staying non-EU nationals, thus, persons that may not be at the location of the border. Another intra-territorialisation example refers to Frontex's conduct of second-line checks on arrivals, namely after the entry and first-line control. Against this backdrop, a discussion has commenced within EU institutions on the prospect of adding reporting for secondary movements in EUROSUR application. If decided, Frontex will start engaging with mobility within the EU. After all, Frontex already includes in its risk analysis reports information about secondary movements (2020a). Hence, Frontex shifts border control accounting simultaneously for extra-territorialisation and intra-territorialisation.

Another border control assumption of Frontex is intelligence. Frontex defines intelligence as 'any information, received or generated, that is related to one of the components of the risk', namely to a threat, vulnerability, or impact (Frontex, 2013a: 15). For Frontex, by collecting, producing, analysing, and then sharing intelligence, the agency maintains border and internal security of the EU, as it enables a better understanding of the operational environment and a more effective management of the external borders (Frontex, 2013a). The importance attributed to intelligence by Frontex is evident in the following institutional discourses:

[Head of the Frontex Situation Centre] 'The future is intelligence-led working'

(Frontex, 2019c).

[Frontex Executive Director] '[...] from the beginning we were extremely clear that we must be an intelligence-driven organisation'

(Frontex, 2010a: 16).

[Frontex Executive Director] 'In a complex world engulfed in a once-in-a-lifetime global crisis [...], reliable intelligence has become an invaluable commodity'

(Frontex, 2021e: 6).

[Frontex officer] 'We need to look at intentions. To the people that have not yet done a criminal act, but have the intention to do so'

(Interviewee 3).

All Frontex operations are intelligence-driven. This means that they are organised based on risk analysis.[6] For example, CIRAM, whilst assessing risks, uses, sorts, processes, produces, and evaluates intelligence. In addition, Frontex has promoted the construction of risk analysis networks and risk analysis cells in third countries for the collection of intelligence. Until now, 7 risk analysis cells are in operation in African countries (Togo, Senegal, Niger, Nigeria, Ghana, Guinea, and The Gambia), whereas there are discussions for the establishment of another 3 cells in Guinea, Côte d'Ivoire, and Mauritania. Also, the agency has decided to start using services for social media analysis to enhance its intelligence (Management Board, 2019), whereas since 2012 it has started promoting artificial intelligence lie-detecting systems. In this context, in 2019, it commissioned RAND Europe to carry out a research study so as to explore the value of the inclusion of artificial intelligence capabilities in border control. Besides, Frontex emphasises its debriefing activities. Debriefing allows Frontex officers to collect directly from irregular border crossers intelligence about crime networks and smuggling routes. Debriefing's prominence for Frontex is attested by the fact that in 2019 the EBCG teams were composed of 696 debriefers compared, for instance, to just 155 stolen vehicles detection officers (Frontex, 2020e). Also, only in 2016 Frontex carried out 3,861 debriefing interviews (Frontex, 2017e: 83). Aside from the collection of intelligence, the agency also shares its intelligence via intelligence products, such as risk analysis and risk assessment reports, alerts, periodic briefings, and situational overviews.

Surveillance constitutes another Frontex border control assumption. According to the Schengen Borders Code, border control involves both border checks and surveillance (Regulation, 2016). It seems, though, that border surveillance has taken over from border checks and now constitutes EU's border control driving force (Jeandesboz, 2011: 117). For Frontex, 'safe, secure and well-functioning external borders are highly dependent on [...] border surveillance' (Frontex, 2021a: 15). Actually, that is one of Frontex's strategic objectives as described in its programming documents. Similarly, pre-frontier wide area surveillance, namely the monitoring beyond the border, constitutes one of Frontex's proposed actions included in its Technical and Operational Strategy for the European Integrated Border Management. In the Frontex Executive Director's words, 'the purpose of border surveillance [...] is to prevent illegal border crossers' (EBCG Day, 2014). Here, border surveillance encloses a proactive context, in which vision is turned into action, when border guards, identifying a group of people or a sea vessel suspected of attempting to cross the border irregularly through the collection of data from visual representation of the border zone, such as radar images, intervene to prevent irregular border entry or intercept them (Dijstelbloem et al., 2017: 226–229). The prominence of surveillance to Frontex operational activities is also reflected in the following discourse:

> [Frontex spokesperson] 'The primary aim of [Mediterranean Frontex sea] operations is to increase surveillance for border control purposes'
> (DW, 2013).

Using an array of sophisticated surveillance devices, such as radars, satellites, remotely piloted aircraft systems, aerostats, and sensors as well as specialised human resources, namely border surveillance officers, the agency seeks to maintain real-time and 24/7 surveillance at the EU external borders and at the pre-frontier area.[7] All the information deriving from surveillance is being collected and sorted by the Frontex Situation Centre. The EUROSUR system, which Frontex operates and coordinates, constitutes the main enabler and implementer of this surveillance environment. This system involves a shared IT platform, which allows participating authorities, namely member states and Frontex, to instantly see, exchange data, and assess the situation at and beyond the EU external border (Regulation, 2013). Currently, EUROSUR is being implemented under the EUROSUR Fusion Services, providing expanded surveillance services with the use of optical and radar satellite technology, including even data from the EU Copernicus earth observation programme.

At the same time, Frontex fosters a technocratic border control. Being an agency, Frontex represents a bureaucratic and managerial solution that brings to the fore its technocratic expertise and authority (Haas, 1992: 11). Actually, Frontex's creation was an attempt to de-politicise EU borders (Neal, 2009). Building on this, Frontex's activities are perceived as neutral and managerial (Paul, 2017: 704), based on the agency's expert technical knowledge (Martins & Jumbert, 2020). Moreover, on its webpage, Frontex is being self-identified as a 'centre of expertise in the area of border control'. Advancing its technocratic function, Frontex's documents regularly evoke the agency's experience and knowledge. Accordingly, an analysis of Frontex's programming documents evidences a constant use of the words: 'efficiency', 'effectiveness' (Horsti, 2012: 303), and 'professional/(s)', which constitute main elements of a technocratic bureaucracy (Centeno, 1993: 311). Another technocratic characteristic stems from Frontex's focus on numbers. It reports on a monthly basis the number of detected irregular border crossings, functioning as a supplier of objective data and neutral facts. To do so, it uses technocratic tools, such as computer systems, standardised templates, and report formats. Furthermore, the agency develops handbooks, best practices, and standard operational procedures. It also periodically evaluates its activities and establishes centres of excellence. It has a management board and an Executive Director. It sets business objectives, business plans, and characterises member states as its customers (Frontex, 2021a), putting forward a managerial perspective. In addition, it manages its operational resources through a management system entitled Opera. So, Frontex clearly implements a technocratic approach to border control. This does not mean, though, that it is exempted from politicisation.

Moving to border control practices, Frontex's work is based on and is materialised through multilateral cooperation. An illustrative example is its joint operations that rely on member states' contribution of officers and equipment. Frontex currently cooperates with 34 national authorities from

European states, such as the Spanish Guardia Civil and the Norwegian Politi. It also collaborates with EU institutions like the European Commission, providing technical expertise, or other EU agencies, such as the European Union Agency for Asylum (EUAA)[8] in hotspot areas, or the European Maritime Safety Agency (EMSA), while implementing its coastguard functions. Frontex's mood for synergy is not restricted to the European soil. It cooperates with international organisations, like NATO and non-EU countries. Indeed, until today, it has concluded Working Arrangements with 18 third countries, including USA, Canada, Russia, and Cape Verde. These documents allow Frontex to build a relationship with third actors by setting the modalities of their cooperation. The same applies to the signature of Memoranda of Understanding, such as with Turkey. This cooperative logic is also manifested discursively:

> [Frontex Executive Director] 'It is in cooperation with the national coast guards that Frontex forms the European Border and Coast Guard, and it is in cooperation that we protect the EU's external borders'
>
> (Frontex, 2018b).

> [Frontex Executive Director] 'Common challenges…require a joint response. We can only effectively tackle many challenges at our borders if we work together across borders'
>
> (Frontex, 2019d).

> [Head of Risk Analysis Unit] 'We had to build a community based on the recognition that I need to share with you what I know so that we could work together'
>
> (Frontex, 2010a: 64).

Another border control practice is professionalisation. Indeed, the agency's webpage lists professionalism as the first value of Frontex, assuring that the agency has 'the knowledge, skills, and competencies needed to fulfil [its] mission' (Frontex, 2022a). Underscoring professionalism, Frontex's reports state that the agency's staff 'share and live the corporate values' and, as a result, 'they perform their activities in a highly professional way' (Frontex, 2014: 12). Hence, Frontex is being self-identified in terms of professionalism. This is also shared by Frontex staff:

> [Head of Frontex's Information Fusion Centre] 'I am happy to lead a talented team of professionals in the fullest sense of the word'
>
> (Frontex, 2022c).

> [Director of Frontex Operations Division] 'I was impressed with the high level of professionalism and expertise of the people working here'
>
> (Frontex, 2010a: 16).

Constructing Frontex's culture

> [Frontex Expert] 'Professionalism and cooperation are the first things that come to my mind when I think about Frontex'
>
> (Frontex, 2022d).

Seeking and urging for professionalism, Frontex has created codes of conduct that set professional and behavioural standards. It has also developed professional education and training, which entail skills and competencies specifically designed for and addressed to border guards, enabling therefore their career advancement. Moreover, Frontex produces programming documents, periodic reports, and handbooks with best practices and operational guidelines on a regular basis. These documents introduce specific rules and technical procedures fostering, in turn, border control's professionalisation. By the same token, they create organisational knowledge (Interviewee 2). These rules and procedures, on the one hand, advance the profession of border guarding, demarcating it from other law-enforcement sectors. On the other hand, they place Frontex at the centre of these processes by being their creator and manager.

Frontex also operates through information gathering. This enables the agency to acquire situational awareness, share data with its partners, and analyse border control trends. The orchestrator of this endeavour is the Frontex Situation Centre, which gathers, sorts out, and then shares all the collected information. Information gathering takes place in almost every Frontex activity, such as the operation of liaison officers, the creation of regional operational offices, and the establishment of networks and partnerships with other actors. To implement and foster information gathering, Frontex has designed and now operates new IT reporting systems, like JORA, FOSS, and EUROSUR. The users of these systems upload border control data or other relevant incidents, such as natural disasters, terrorist attacks, border conflicts, or even contagious diseases among Frontex staff. The importance of information gathering for Frontex is also traced in the following discourses:

> [Frontex Executive Director] 'Frontex will strive to strengthen data and information collection'
>
> (Frontex, 2021a: 4).

> [Frontex officer] 'To take an operational decision we rely on risk analysis and information from networks'
>
> (Interviewee 1).

It seems that border control has entered an 'Information Age' (Frontex, 2013b). Frontex's function enables this information environment with the sharing of data. Yet, simultaneously, this information environment strengthens Frontex's role, because of the agency's privileged position to manage and analyse all the information and knowledge (Haas, 1992; Scott, 2008).

Regarding technology, the agency employs modern technological tools to perform its tasks, such as IT platforms, automated tracking software, geospatial imagery, space-based infrastructure, remotely piloted aircraft systems, artificial intelligence, and virtual reality. To develop and use state-of-the-art technology it collaborates with the industry sector, and especially defence and surveillance companies (Marin, 2011; Csernatoni, 2018). For this, central is the function of the Research and Innovation Unit, which, according to Frontex's own wording, 'is rapidly becoming the source for member states needing advice on new technology' (Frontex, 2010a: 55). As a result, Frontex's role is dual. It can become the initiator of a research project or technology, requesting the industry to search for and propose solutions to specific needs identified by itself or its member states. By contrast, it can function as the buyer, tester or end-user of new technological products already developed by the industry, which it then deploys at the EU external border. Yet, the inclusion of these technological tools in the border control conduct fundamentally influences and eventually alters border control's nature. Arguably, each new technological tool deployed reconfigures both the space as a territoriality and the social landscape, creating new realities and meanings regarding borders and border control (Walters, 2006).

The last border control practice is policing. The processes of persons' identification, authentication, and filtering that Frontex regularly performs constitute part of a policing methodology (Dijstelbloem et al., 2017: 228). This methodology involves the use of reporting systems, which graphically visualise each border control incident with specific details regarding the location, time, and means involved. Moreover, Frontex collects background information about individuals and their actions, producing, processing, and managing knowledge about persons (Tazzioli & Walters, 2016: 454; Campesi, 2022: 198–199). This is being implemented, for instance, with debriefing, which constitutes a form of police interrogation, as well as lie-detection technology. These activities render Frontex a border-policing agency (Aas & Gundhus, 2015; Pallister-Wilkins, 2015). After all, most Frontex staff have a policing background, taking into account that experience in law enforcement or police is often an essential prerequisite for Frontex recruitment as proven by Frontex job vacancies published in its institutional webpage. Thus, they operate and think in line with policing terms. This is evident in Frontex's documents. For instance, Frontex's incident reports are characterised by a factual and evidence-based writing referring solely to the time and location of events, with clear and neutral language, like 'this morning at 6:45, a Royal Netherlands Marechaussee patrol boat [...] detected [...]' (Frontex, 2019e). Also, there is regular reference to 'crime' and 'criminal'. This wording and method of writing are usually encountered in police documents. In this policing spirit, there are document versions for 'law-enforcement only', whereas many operational documents are not publicly accessible. At the same time, Frontex has been criticised in academia for maintaining a level of secrecy (Carrera, 2007; Pollak & Slominski, 2009), withholding information and shielding the content of its management board decisions from the public. Also, rather often it restructures its official webpage, rendering more difficult than before the navigation for the extraction of data. For instance, currently, an external user has to visit at least two different sections to find and download Frontex documents, whereas before all published Frontex documents were under one title. There are also important amendments in the published content. This concerns the section of operations, given that a user was able to see and enter each Frontex operation conducted and not just the main operations, as it is now. Furthermore, for an 'outsider', contacting Frontex staff directly can become a real challenge or even an insurmountable task, due to institutional 'gatekeepers', which, compared to other EU agencies or EU institutions, seem keen on preserving Frontex's distancing from the public as well as filtering and authorising whatever will be circulated to the outside. The above restrictions come across usually in policing environments.

These assumptions and practices compose Frontex's cultural traits, given that they dominate the agency's discourse and action, as shown from the above analysis. Both the staff of Frontex as well as its 'outsiders' consciously or subconsciously link Frontex with these elements. They refer, for instance, to logics of securitisation and re-territorialisation, to its surveillance context and technocratic spirit, to the deployment of technological tools as well as to Frontex's policing methods. All these elements shape this EU agency and feed our conception about what is Frontex and its role in border control (see Table 3.1).

Table 3.1 Frontex's border control cultural traits

Border assumptions	EU external border
	National territory
Border control assumptions	Fundamental rights
	Securitisation
	Extra-territorialisation/
	Intra-territorialisation
	Technocracy
	Surveillance
	Intelligence
Border control practices	Policing
	Information gathering
	Multilateral cooperation
	Technology
	Professionalisation

Conclusion

This chapter inserted culture into Frontex's analysis. The aim was to show that Frontex, despite being an EU agency, it is also an actor capable of developing and pursuing its own culture. Indeed, the analysis showed that Frontex has developed cultural traits, which are composed of specific border and border control assumptions and practices.

The chapter firstly explored Frontex as a border control actor, arguing that it does not just implement but, rather, formulates border policy. In this spirit, it manages human and technical resources, creates new structures, develops knowledge, collects and analyses data, assesses risks and capabilities, manages crises, creates social relations, has a vision, as well as sets goals and articulates strategic objectives. With its presence, action, and involvement in border control, Frontex has gained acceptance and recognition by its partners, namely member states, EU institutions, and third countries or other organisations. This means that Frontex is an actor, which acts at its social environment interacting with other actors.

Then, the chapter moved to Frontex's cultural traits. Applying the 'cultures of border control' analytical framework (Zaiotti, 2011), it analysed Frontex's border assumptions as well as its border control assumptions and practices, which constitute its cultural traits. Analysing these assumptions and practices, the chapter showed that they reflect the workings of this agency and construct Frontex's border control role. Actually, they are part of its institutional culture as they are systematically promoted by the agency, dominating both its actions and discourses. In parallel, these assumptions and practices inform the border control conduct, as Frontex, being a border control actor, constitutes key part of it. Yet, it still remains unanswered whether these elements compose a border control culture currently pursued. To answer this, I turn to borders.

Notes

1 See, for instance, European Commission (2022).
2 See, for instance, Albanian PM's Office (2018) and Cyprus Mail (2021).
3 For more information on Frontex-UK relation, see Chapter 7.
4 See, for instance, Frontex (2010a).
5 For the conception of 'borderlands', see Del Sarto (2010).
6 For moral issues linked to Frontex's use and production of intelligence, see Gundhus (2017).
7 For challenges in real-time data circulation, see Pollozek (2020).
8 Formerly known as EASO.

References

Aas, K. & Gundhus, H. (2015) 'Policing Humanitarian Borderlands: Frontex, Human Rights and the Precariousness of Life'. *The British Journal of Criminology*, 55(1), pp. 1–18.
Albanian PM's Office (2018) 'Joint Press Conference of Prime Minister Edi Rama, Commissioner Dimitris Avramopoulos, and the European Border and Coast Guard

Agency Executive Director Fabrice Leggeri'. Available at https://kryeministria. al/en/newsroom/operacioni-i-frontex-thellim-i-bashkepunimit-me-be-ne-fushen-e-sigurise/ (accessed January 2022).

Allen, D. & Smith, M. (1990) 'Western Europe's Presence in the Contemporary International Arena'. *Review of International Studies*, 16(1), pp. 19–37.

Avramopoulos, D. (2015) 'Speech by Commissioner Avramopoulos'. Available at https://avramopoulos.gr/en/content/speech-commissioner-avramopoulos-frontex-conference-european-day-border-guards-warsaw-poland (accessed January 2022).

Berger, T. (2000) 'Set for Stability? Prospects for Conflict and Cooperation in East Asia'. *Review of International Studies*, 26(3), pp. 405–428.

Börzel, T.A. & Risse, T. (2003) 'Conceptualizing the Domestic Impact of Europe'. In: K. Featherstone & C. Radaelli (eds.), *The Politics of Europeanization*. Oxford: Oxford University Press, pp. 57–82.

Buzan, B., Wæver, O. & de Wilde, J. (1998) *Security: A New Framework for Analysis*. Boulder: Lynne Rienner.

Campesi, G. (2022) *Policing Mobility Regimes: Policing Mobility Regimes*. Abingdon: Routledge.

Carrera, S. (2007) 'The EU Border Management Strategy: Frontex and the Challenges of Irregular Migration in the Canary Islands'. *CEPS*, Working Document 261, pp. 1–33.

Centeno, M.A. (1993) 'The New Leviathan: The Dynamics and Limits of Technocracy'. *Theory and Society*, 22(3), pp. 307–335.

Chillaud, M. (2012) 'Frontex as the Institutional Reification of the Link between Security, Migration and Border Management'. *Contemporary European Studies*, 2, pp. 45–61.

Copeland, D. (2000) 'The Constructivist Challenge to Structural Realism: A Review Essay'. *International Security*, 25(2), pp. 187–212.

Csernatoni, R. (2018) 'Constructing the EU's High-Tech Borders: FRONTEX and Dual-Use Drones for Border Management'. *European Security*, 27(2), pp. 175–200.

Cyprus Mail (2021) 'Cyprus Asks Frontex to Stop Migrants Sailing to North'. Available at https://cyprus-mail.com/2021/04/10/cyprus-asks-frontex-to-stop-migrants-sailing-to-north/ (accessed January 2022).

Del Sarto, R. (2010) 'Borderlands: The Middle East and North Africa as the EU's Southern Buffer Zone'. In: D. Bechev & K. Nicolaïdis (eds.), *Mediterranean Frontiers: Borders, Conflicts and Memory in a Transnational World*. London: I.B. Tauris, pp. 149–167.

Dijstelbloem, H., van Reekum, R. & Schinkel, W. (2017) 'Surveillance at Sea: The Transactional Politics of Border Control in the Aegean'. *Security Dialogue*, 48(3), pp. 224–240.

DW (2013) 'Migration is a Phenomenally Complex Issue'. Available at https://www.dw.com/en/migration-is-a-phenomenally-complex-issue/a-17152434 (accessed January 2022).

EBCG Day (2014) 'ED4BG 2014 Integrated Border Management and its Way Forward'. Available at https://www.youtube.com/watch?v=MAydhOJlRns (accessed January 2022).

EEAS (2016) 'European Commission and Frontex Assess Implementation of EU Assistance'. Available at https://eeas.europa.eu/headquarters/headquarters-homepage/13084/europeancommission-and-frontex-assess-implementation-eu-assistance_en (accessed January 2022).

EUobserver (2020) 'Thousands Apply for EU Border Guard Posts'. Available at https://euobserver.com/justice/147186 (accessed February 2022).

EURACTIV (2021) 'EU to Detail Future Relationship with Gibraltar Next Week'. Available at https://www.euractiv.com/section/uk-europe/news/eu-to-detail-future-relationship-with-gibraltar-next-week/ (accessed January 2022).

European Commission (2020) 'Report: First Multiannual Evaluation Programme (2015–2019)'. COM(2020)779.

European Commission (2022) 'Statistics on Migration to Europe'. Available at https://ec.europa.eu/info/strategy/priorities-2019-2024/promoting-our-european-way-life/statistics-migration-europe_en (accessed January 2022).

European Court of Auditors (2020) *Audit Preview: Frontex*. Luxembourg: ECA.

Ferrer-Gallardo, X. & van Houtum, H. (2014) 'The Deadly EU Border Control'. *ACME: An International E-Journal for Critical Geographies*, 13(2), pp. 295–304.

Finnemore, M. (1996) *National Interests in International Society*. Ithaca: Cornell University Press.

Frontex (2007) 'Frontex's Patronage over the Conference Freedom and Security: Europe Without Borders'. Available at https://frontex.europa.eu/media-centre/news/news-release/frontex-s-patronage-over-the-conference-freedom-and-security-europe-without-borders–n7lew3 (accessed January 2022).

Frontex (2010a) *Beyond the Frontiers*. Warsaw: Frontex.

Frontex (2010b) 'Frontex Operational Office Opens in Piraeus'. Available at https://frontex.europa.eu/media-centre/news/news-release/frontex-operational-office-opens-in-piraeus-hk4q3Z (accessed January 2022).

Frontex (2010c) 'Papoutsis, Besson, Malmström and Laitinen Visit RABIT Operational Area'. Available at https://frontex.europa.eu/media-centre/news/news-release/papoutsis-besson-malmstrom-and-laitinen-visit-rabit-operational-area-jkKZi9 (accessed January 2022).

Frontex (2012) 'Frontex Responds to the European Ombudsman'. Available at https://frontex.europa.eu/media-centre/news/news-release/frontex-responds-to-the-european-ombudsman-8oCMDO (accessed January 2022).

Frontex (2013a) *Common Integrated Risk Analysis Model*. Warsaw: Frontex.

Frontex (2013b) 'Border Control in the Information Age'. Available at https://frontex.europa.eu/media-centre/news/focus/border-control-in-the-information-age-udh57L (accessed January 2022).

Frontex (2014) *Frontex Programme of Work 2015*. Warsaw: Frontex.

Frontex (2017a) 'Better Knowledge Is Better Management'. Available at https://frontex.europa.eu/media-centre/news/focus/better-knowledge-is-better-management-first-european-joint-master-s-in-strategic-border-management-KyMj8P (accessed January 2022).

Frontex (2017b) 'Frontex and Interpol Hold First Joint Conference on Document Fraud'. Available at https://frontex.europa.eu/media-centre/news/news-release/frontex-and-interpol-hold-first-joint-conference-on-document-fraud-9uuJsy (accessed January 2022).

Frontex (2017c) 'EU Commissioner King Visits Frontex'. Available at https://frontex.europa.eu/media-centre/news/news-release/eu-commissioner-king-visits-frontex-IVmKXq (accessed January 2022).

Frontex (2017d) *Risk Analysis for 2017*. Warsaw: Risk Analysis Unit.

Frontex (2017e) *Annual Activity Report 2016*. Warsaw: Frontex.

Frontex (2018a) 'Frontex Marks Two Years as the European Border and Coast Guard Agency'. Available at https://frontex.europa.eu/media-centre/news/news-release/frontex-marks-two-years-as-the-european-border-and-coast-guard-agency-ECWley (accessed January 2022).
Frontex (2018b) 'Frontex Hosts EBCG Day to Discuss Security at Sea'. Available at https://frontex.europa.eu/media-centre/news/news-release/frontex-hosts-european-border-and-coast-guard-day-to-discuss-security-at-sea-vb539m (accessed January 2022).
Frontex (2019a) 'Frontex Hosts EBCG Day'. Available at https://frontex.europa.eu/media-centre/news/news-release/frontex-hosts-european-border-and-coast-guard-day-eJnxSD (accessed January 2022).
Frontex (2019b) 'Become a Trainee at EU's Most Dynamic Agency'. Available at https://frontex.europa.eu/media-centre/news/news-release/become-a-trainee-at-eu-s-most-dynamic-agency-1p07gA (accessed June 2022).
Frontex (2019c) 'EBCG Day 2019'. Available at https://www.youtube.com/watch?v=Xv2hUOVYJ2I (accessed January 2022).
Frontex (2019d) 'Frontex Management Board Meets with Partners from the Western Balkans'. Available at https://frontex.europa.eu/media-centre/news/news-release/frontex-management-board-meets-with-partners-from-the-western-balkans-1wf8k8 (accessed January 2022).
Frontex (2019e) 'European Border and Coast Guard Agency (Frontex) Involved in Search and Rescue Operation off Lesvos'. Available at https://frontex.europa.eu/media-centre/news/news-release/european-border-and-coast-guard-agency-frontex-involved-in-search-and-rescue-operation-off-lesvos-Q3sC4e (accessed January 2022).
Frontex (2020a) *Risk Analysis for 2020*. Warsaw: Risk Analysis Unit.
Frontex (2020b) *Consolidated Annual Activity Report 2019*. Warsaw: Frontex.
Frontex (2020c) 'Frontex to Provide Border Security Expertise to European Commission's Research Projects'. Available at https://frontex.europa.eu/media-centre/news/news-release/frontex-to-provide-border-security-expertise-to-european-commission-s-research-projects-ZrCBoM (accessed January 2022).
Frontex (2020d) 'Frontex Launches Rapid Border Intervention on Greek Land Border'. Available at https://frontex.europa.eu/media-centre/news/news-release/frontex-launches-rapid-border-intervention-on-greek-land-border-J7k2lh (accessed January 2022).
Frontex (2020e) *Report on the Operational Resources in 2019*. Warsaw: Frontex.
Frontex (2021a) *Single Programming Document 2021-2023*. Warsaw: Frontex.
Frontex (2021b) *Consolidated Annual Activity Report 2020*. Warsaw: Frontex.
Frontex (2021c) *Fundamental Rights Strategy*. Warsaw: Frontex.
Frontex (2021d) *Annual Report: Fundamental Rights Officer 2020*. Warsaw: Frontex.
Frontex (2021e) *Risk Analysis for 2021*. Warsaw: Risk Analysis Unit.
Frontex (2022a) 'Our Mission'. Available at https://frontex.europa.eu/about-frontex/our-mission/ (accessed February 2022).
Frontex (2022b) 'FAQ'. Available at https://frontex.europa.eu/about-frontex/faq/frontex-operations/ (accessed February 2022).
Frontex (2022c) 'Our Team: Piotr'. Available at https://frontex.europa.eu/careers/who-we-are/our-team/piotr-from-poland-ngK002 (accessed February 2022).
Frontex (2022d) 'Our Team: Nikolay'. Available at https://frontex.europa.eu/careers/who-we-are/our-team/nikolay-from-bulgaria-oETBZX (accessed February 2022).

Gundhus, H. (2017) 'Negotiating Risks and Threats: Securing the Border through the Lens of Intelligence'. In: N. Fyfe, H. Gundhusn & K. Vrist Rønn (eds.), *Moral Issues in Intelligence-Led Policing*. Abingdon: Routledge, pp. 223245.

Haas, P.M. (1992) 'Introduction: Epistemic Communities and International Policy Coordination'. *International Organization*, 46(1), pp. 1–35.

Horii, S. (2016) 'The Effect of Frontex's Risk Analysis on the European Border Controls'. *European Politics and Society*, 17(2), pp. 242–258.

Horsti, K. (2012) 'Humanitarian Discourse Legitimating Migration Control: FRONTEX Public Communication'. In: M. Messier, R. Schroeder & R. Wodak (eds.), *Migrations: Interdisciplinary Perspectives*. Vienna: Springer, pp. 297–308.

House of Lords (2017) 'Leaving the European Union: Frontex and UK Border Security Cooperation Within Europe'. *In Focus*, LIF 2017/0039, pp 1–4.

Jeandesboz, J. (2011) 'EUROSUR and the Ethics of European Border Control Practices'. In: P. Burgess & S. Gutwirth (eds.), *A Threat against Europe? Security Migration and Integration*. Brussels: Brussels University Press, pp. 11–132.

Kingdon, J.W. (2014 [1984]) *Agendas, Alternatives, and Public Policies*. Essex: Pearson Education Limited.

Lægreid, P., Roness, P. & Verhoest, K. (2011) 'Explaining the Innovative Culture and Activities of State Agencies'. *Organization Studies*, 32(10), pp. 1321–1347.

Léonard, S. & Kaunert, C. (2022) 'The Securitisation of Migration in the European Union: Frontex and Its Evolving Security Practices'. *Journal of Ethnic and Migration Studies*, 48(6), pp. 1417–1429.

Léonard, S. (2010) 'EU Border Security and Migration into the European Union: Frontex and Securitisation through Practices'. *European Security*, 19(2), pp. 231–254.

Management Board (2019) Frontex Management Board Decision 8/2019 Adopting Amendment N1 to the Programming Document 2019–2021.

Marin, L. (2011) 'Is Europe Turning into a "Technological Fortress"? Innovation and Technology for the Management of EU's External Borders: Reflections on FRONTEX and EUROSUR'. In: M.A. Heldeweg & E. Kica (eds.), *Regulating Technological Innovation: Legal and Economic Regulation of Technological Innovation*. Basingstoke: Palgrave Macmillan, pp. 131–151.

Martins, B.O. & Jumbert, M.G. (2020) 'EU Border Technologies and the Co-Production of Security Problems and Solutions'. *Journal of Ethnic and Migration Studies*. https://doi.org/10.1080/1369183X.2020.1851470 (published online).

NATO (2016) 'Joint Press Point by NATO Secretary General Jens Stoltenberg and the President of the European Commission, Jean-Claude Juncker'. Available at http://www.nato.int/cps/en/natohq/opinions_129162.htm?selectedLocale=en (accessed January 2022).

Neal, A. (2009) 'Securitization and Risk at the EU Border: The Origins of FRONTEX'. *Journal of Common Market Studies*, 47(2), pp. 333–356.

Pallister-Wilkins, P. (2015) 'The Humanitarian Politics of European Border Policing: Frontex and Border Police in Evros'. *International Political Sociology*, 9(1), pp. 53–69.

Pascouau, Y. & Schumacher, P. (2014) 'Frontex and the Respect of Fundamental Rights: From Better Protection to Full Responsibility'. *EPC*, Policy Brief, pp. 1–4.

Paul, R. (2017) 'Harmonisation by Risk Analysis? Frontex and the Risk-Based Governance of European Border Control'. *Journal of European Integration*, 39(6), pp. 689–706.

Perkins, C. & Rumford, C. (2013) 'The Politics of (Un)fixity and the Vernacularization of Borders'. *Global Society*, 27(3), pp. 267–282.
Perkowski, N. (2018) 'Frontex and the Convergence of Humanitarianism, Human Rights and Security'. *Security Dialogue*, 49(6), pp. 457–475.
Pollak, J. & Slominski, P. (2009) 'Experimentalist but Not Accountable Governance? The Role of Frontex in Managing the EU's External Borders'. *West European Politics*, 32(5), pp. 904–924.
Pollozek, S. (2020) 'Turbulences of Speeding Up Data Circulation: Frontex and Its Crooked Temporalities of "Real-Time" Border Control'. *Mobilities*, 15(5), pp. 677–693.
Regulation (EU) 2016/399 of the European Parliament and of the Council of 9 March 2016 on a Union Code on the Rules Governing the Movement of Persons across Borders (Schengen Borders Code) [2016, OJ L 77/1].
Regulation (EU) 2019/1896 of the European Parliament and of the Council of 13 November 2019 on the European Border and Coast Guard and Repealing Regulations (EU) No 1052/2013 and (EU) 2016/1624 [2019, OJ L 295/1].
Regulation (EU) No 1052/2013 of the European Parliament and of the Council of 22 October 2013 Establishing the European Border Surveillance System (Eurosur) [2013, OJ L 295/11].
Reid-Henry, S.M. (2013) 'An Incorporating Geopolitics: Frontex and the Geopolitical Rationalities of the European Border'. *Geopolitics*, 18(1), pp. 198–224.
Rijpma, J. (2012) 'Hybrid Agentification in the Area of Freedom, Security and Justice and Its Inherent Tensions: The Case of Frontex'. In: M. Busuioc, M. Groenleer & J. Trondal (eds.), *The Agency Phenomenon in the European Union: Emergence, Institutionalisation and Everyday Decision-Making*. Manchester: Manchester University Press, pp. 84–102.
Rijpma, J. (2016) 'The Proposal for a European Border and Coast Guard: Evolution or Revolution in External Border Management?'. *LIBE Committee*, Study, pp. 1–36.
Saurugger, S. (2013) 'Constructivism and Public Policy Approaches in the EU: From Ideas to Power Games'. *Journal of European Public Policy*, 20(6), pp. 888–906.
Schimmelfening, F. (2003) *The EU, NATO and the Integration of Europe: Rules and Rhetoric*. Cambridge: Cambridge University Press.
Scott, W.R. (2008) *Institutions and Organizations: Ideas and Interests*. London: SAGE.
Searle, J.R. (1995) *The Construction of Social Reality*. London: Penguin.
Tazzioli, M. & Walters, W. (2016) 'The Sight of Migration: Governmentality, Visibility and Europe's Contested Borders'. *Global Society*, 30(3), pp. 445–464.
Vaughan-Williams, N. (2008) 'Borderwork beyond Inside/Outside? Frontex, the Citizen-Detective and the War on Terror'. *Space and Polity*, 12(1), pp. 63–79.
Vollmer, R. & von Boemcken, M. (2014) 'Europe off Limits: Militarization of the EU's External Borders'. In: I. Werkner, J. Kursawe, M. Johannsen, B. Schoch & M. von Boemcken (eds.), *Friedensgutachten 2014*. Münster: LIT, pp. 57–69.
Walters, W. (2006) 'Rethinking Borders Beyond the State'. *Comparative European Politics*, 4(2/3), pp. 141–159.
Wiermans, M. (2012) *The Securitization of Frontex: A Discours Analysis*. Saarbrücken: Lambert.
Zaiotti, R. (2011) *Cultures of Border Control: Schengen & the Evolution of European Frontiers*. Chicago: University of Chicago Press.

Zürn, M. & Checkel, J.T. (2005) 'Getting Socialized to Build Bridges: Constructivism and Rationalism, Europe and the Nation-State'. *International Organization*, 59(4), pp. 1045–1079.

Interviews

Interviewee 1: Frontex officer in Warsaw, 11 June 2018.
Interviewee 2: Frontex officer in Warsaw, 12 June 2018.
Interviewee 3: Frontex officer in Warsaw, 13 June 2018.
Interviewee 7: Border guard in southern Evros, 23 October 2018.

4 Constructing the border of Lampedusa

'The challenges that Lampedusa and Italy
are facing are European challenges'.
—José Manuel Barroso, former European Commission President
(European Commission, 2013)

Introduction

On hearing the term 'border control', the next thing that comes to mind is 'borders'. Any analysis of border control cannot avoid the close examination of the actual borders. Borders like Lampedusa: a small yet picturesque Italian island in the Mediterranean Sea, located between Sicily and the African coast. In the last decades, Lampedusa has been put into the spotlight due to functioning as a door, like its symbolic 'Porta d' Europa'-shaped monument, for vast flows of irregular border crossers trying to reach it and therefore reach Europe. This sea journey, though, incurs a massive death toll, as each year thousands of migrants are drowned, rendering it the deadliest migratory route in the world. This means that Lampedusa is more than just a border.

Indeed, borders are not mere geographic spaces or fixed lines drawn on political maps (Newman, 2006: 175) delineating the geographic location where physical border control takes place and, therefore, manifesting us/them territorial separations. Instead, they enclose a dynamic nature. They are being made (van Houtum et al., 2005; Newman, 2011) to function as sites of power (Wilson & Donnan, 1998: 10), which implement and reproduce border control choices, and power relations among variant actors that are present at the border, such as border guards and border crossers. So, any border control investigation cannot overpass borders. Neither can it disregard what and who surrounds, situates, produces, and governs them.

To examine border control, the journey to the border cannot be avoided. This journey, figuratively and literally, starts with the border of Lampedusa. The goal is to shed light on the border control conduct and border control actors that form this EU external border in order to extract the cultural traits that

DOI: 10.4324/9781003230250-4

construct it. Traits are the empirical traces of the border control culture pursued by the members of the border control community (Zaiotti, 2011: 38). Therefore, these traits are indicators of a border control culture at Europe's borders.

This chapter forms part of a comparative case study analysis that proceeds with the scrutiny of Lampedusa and continues with Evros (Chapter 5). Its aim is to draw parallels and differences between two-variant EU external borders: the Italian maritime border of Lampedusa and the Greek land border of Evros. These parallels and differences allow extracting any common cultural traits for the border control conduct in the form of border control assumptions and practices. Furthermore, they permit depicting Frontex's distinct role in EU border control.

Having Lampedusa as a starting point for borders' exploration, this chapter looks at the border, the border control conduct, the actors involved in border control, including Frontex, as well as the cultural traits encountered at the border, which refer to certain border control assumptions and practices. The analysis relies mostly on data derived from in-situ analysis with fieldwork in Lampedusa and interviews with Italian officers conducted in the island. The argument advanced is that there is a detectable evolution of border control in Lampedusa due to the border's dual nature as a national and EU external border. This is manifested in the assumptions and practices that both construct and govern it. In parallel, the chapter empirically demonstrates that Frontex is a border control actor involved in the border's everyday ordering and bordering (Popescu, 2012), namely the border's management and construction.

The chapter first refers to Lampedusa's border characteristics. Then, it investigates the border control conduct and the actors involved in border control. The next part focuses on Frontex's presence and operational activities there. The last section analyses the cultural traits traced in Lampedusa, namely the border control assumptions and practices derived from the mode of the border control conduct.

The island setting

Lampedusa is Italy's southernmost part. It is a small island of the Mediterranean Sea and capital of Sicily's Pelagie Islands. Morphologically, it is a flat rocky island of 20 square kilometres (km^2). On climate, harsh wind prevails throughout the year, especially during the night. Its approximately 6,000 inhabitants, living mainly in its only town, also called Lampedusa, depend on fishing, agriculture, and tourism, as the island has become a tourist attraction mostly owing to its wildlife fauna, such as migratory birds and Caretta Caretta turtles.

Albeit on Italian soil, Lampedusa is considered a fragment of Africa in Europe (Kitagawa, 2011: 201), because of its geological connection with the African continent by an undersea shelf and its geographic proximity with the shores of North Africa (see Map 4.1). It is situated 113 km from Tunisia, 200 km from Sicily, and 300 km from Libya. This geographic marginality

Constructing the border of Lampedusa 69

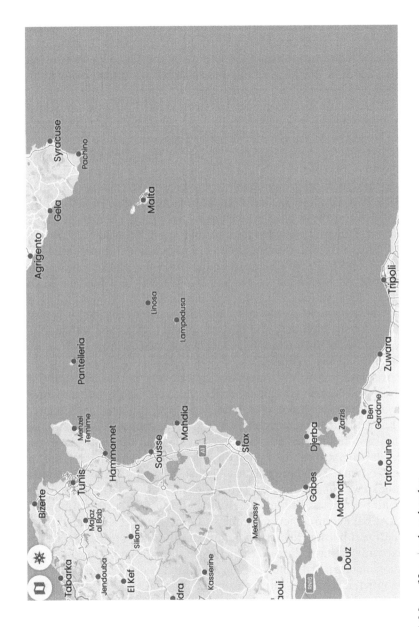

Map 4.1 Map of Lampedusa border.
Source: ©OpenStreetMap, Mapcarta and Mapbox.

from the Italian peninsula means no easy access. Indeed, the island is connected to a few Italian cities by sporadic air flights. Alternatively, by sea, there are 8-hour ferry routes from the Sicilian port of Porto Empedocle.

In spite of its isolation from Italy, Lampedusa enjoys great geostrategic importance, as it is located in the middle of the Sicilian Strait. Its geographic position, though, has incurred various traumatic events in Lampedusa, such as North African pirate attacks, bombing by the Allied Powers during World War II, and in 1986, a scud missiles strike by Muammar al-Qaddafi in retaliation for USA bombing against Libya.

Overall, Lampedusa's development has been shaped by Italy's relations with its southern neighbours. Historically, Ancient Rome's rule over countries of the Northern Africa established social, economic, and political ties. These ties became even more palpable during the mass Italian emigration towards North Africa of the 19th and 20th centuries. For instance, in 1924, amidst fascism in Italy, approximately 91,000 Italians were living in Tunisia (Cresti, 2008: 194). Yet, Italy's colonial aspirations and competition for influence in the Mediterranean basin as well as its wish for participation in the trans-Saharan trade routes fuelled tensions in the neighbourhood. France's invasion of Tunisia in 1881, though the Italian population outnumbered the French, put an end to Rome's plans to render Tunisia an Italian protectorate. To continue exerting influence over the country, Italy started to build an Italo-Tunisian cooperation based on economic incentives. However, Italy's wish for a Mediterranean colony materialised when Italy invaded Libya in 1911 after one of the bloodiest and fiercest colonial wars. The Italian rule over Libya ceased in 1943, when Italy lost World War II. Yet, Italy's economic influence in Libya remained.

The above description of Lampedusa describes a porous relationship between Italy and North African countries, characterised by cross-border exchanges, development links, population mobility, and even neighbourly tensions. Lampedusa, being located between Libya, Tunisia, Malta, and the Italian mainland, functioned as a mobility corridor and a small trade centre (Li Causi, 1987: 165). Actually, Lampedusa became a trans-Mediterranean channel embodying the heterogeneous elements that unite and divide the Mediterranean region and the people that live on its shores. Yet, in the last decades, Lampedusa has become reconstructed as an EU external border. As a result, it has acquired new meaning and function that was added to its role as a national border. Therefore, Lampedusa, on the one hand, constitutes a place advancing a Euro-African link built on cross-cultural merges and common memories (Brambilla, 2016: 116). On the other hand, as an EU external frontier, it separates Europe from the 'unwanted' African migrants (van Houtum & Mamadouh, 2008: 95).

The making of a border

Being an island, mobility towards and from Lampedusa mainly depends on the sea route, with ferries and ships arriving at its two seaports: the new port and the old port. Most ferries arriving on Lampedusa use the domestic sea route departing from Sicily. So, passengers arriving at the island are not normally subject to passport control. That said, a typical border control process for regular sea arrivals entails travel document control before embarkation and after disembarkation from the ferry. Besides formal mobility for tourism or commerce, in the last decades, Lampedusa has become a primary entry point for irregular migrants from Africa trying to reach Europe using the Central Mediterranean route.[1]

In the 1990s, Lampedusa became a destination for many irregular border crossers, mainly from Tunisia, who were arriving in Italy to find seasonal work in the fishing or agricultural sector (Global Initiative, 2014). Even so, those flows were not preoccupying Italians, as the country's focus was on irregular migratory movements from the Balkans, especially Albanian immigrants reaching the region of Puglia via the Adriatic Sea and the Strait of Otranto.

In the 2000s, mobility altered significantly both in nature and size. Lampedusa soon became the main entry point for the majority of irregular flows to Italy. For instance, in 2006, 18,047 irregular border crossers reached Lampedusa, whereas the overall number for irregular sea arrivals in Italy was 22,016. Yet, these irregular border crossers did not intend to stay on the island. They were using Lampedusa as a stepping-stone (Bernardie-Tahir & Schmoll, 2014: 90) on their way towards the Italian mainland and then to other north European countries, like Austria, Switzerland, or France (IOM, 2018: 19). To halt the migratory wave towards Lampedusa's shores, Italy and Libya put their bloody colonial past aside, signing a bilateral agreement, ratified in 2009, which enabled them to cooperate against irregular border crossings. This, however, resulted in push-back practices[2] and Italy's condemnation by the European Court of Human Rights (ECtHR, 2012).

The situation deteriorated in 2011. Lampedusa faced a major migratory influx in the wake of the Jasmine Revolution in Tunisia, civil conflict in Libya leading to Qaddafi's death, and the Arab Spring revolts.[3] The peak in sea arrivals amounted to 51,922 irregular border crossers in 2011 compared to just 459 in 2010 (see Table 4.1). Most were Tunisian and Libyan males fleeing from violent uprisings and economic recession. Soon, besides migrant smuggling, the Central Mediterranean route became a corridor for various cross-border crimes, like drug smuggling, namely heroin and cocaine, as well as arms trafficking (Shaw & Mangan, 2014).

The migratory wave towards Lampedusa continued, yet with less intensity compared to the overall irregular migration flow towards Italy (see Chart 4.1). For instance, in 2017, 119,445 irregular border crossers entered the country, of which only 9,057 disembarked in Lampedusa. This considerable reduction in the number of disembarkations in Lampedusa was attributed to various reasons, such as new smuggling modi operandi relating to the hotspot function. Irregular border crossers that were not eligible for relocation became trapped in the hotspot centres without being able to move informally to northern Europe.[4] In fact, some of them refused to be

Table 4.1 Irregular sea migration to Lampedusa

Year	Irregular Border Crossings	Top Nationalities	Causes
2008	31,311	Tunisians, Nigerians	• Tunisia uprising
2009	02,947	Somalians, Nigerians, Tunisians	• Italy-Libya cooperation • Italian anti-immigration legislation
2010	00,459	Tunisians, Algerians	• Italy-Libya agreement
2011	51,922	Tunisians, Libyans	• Jasmine Revolution • Arab Spring • Libyan civil war & military intervention • Emergency state
2012	05,202	Eritreans, Tunisians	• Spanish readmission agreements • Italy-Tunisia agreement
2013	14,753	Eritreans, Syrians	• Border controls in Israel and the Gulf states diverting migration flows to Europe • Saudi Arabia/Eritrea border fence
2014	04,194		• Mare Nostrum • Turmoil in Iraq, Syria, Central African Republic, South Sudan, Eritrea • Lampedusa unsafe harbour, closed detention centre
2015	21,160	Nigerians, Eritreans	• Migration crisis • Hotspot operation • Shift towards Eastern Mediterranean sea route • Operation Sophia
2016	11,557		
2017	09,057	Nigerians, Guineans	• Less Libyan departures
2018	03,468	Tunisians, Eritreans	• New Italian government • Shift towards Spain
2019	4,802	Tunisians, Libyans	
2020	20,685	Tunisians, Bangladeshis	• COVID-19 pandemic • Shift from the Eastern Mediterranean sea route
2021	35,083	Tunisians, Egyptians, Bangladeshis	• New Italian government • Flows towards the Calabrian route (from Turkey to Calabria)

Notes: Author's elaboration of data.
Data for flows extracted from CoE (2011), WHO (2012), Global Initiative (2014), Guardia Costiera (2022), and Ministero dell'Interno (2022). Data for 2020 and 2021 were provided by the Italian Ministry of Interior (Ministero dell'Interno, Direzione Centrale Immigrazione e Polizia delle Frontiere) on 15 April 2022 in the form of a written reply after an official request made by the author.

Constructing the border of Lampedusa 73

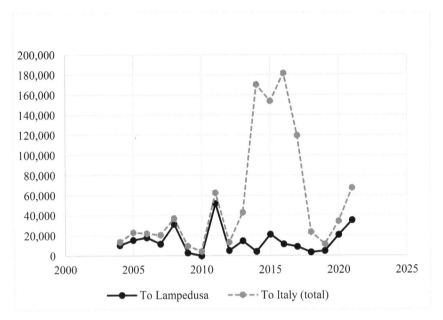

Chart 4.1 Irregular sea arrivals to Italy.

fingerprinted to avoid detection of subsequent intra-Schengen movements (Tazzioli, 2018: 2770). Taking into account that the first Italian hotspot was established in September 2015 in Lampedusa, this may have affected the smuggling routes and particularly the disembarkation place, with migrants opting to land instead somewhere else.

Disembarkation, though, is rarely a matter of migrants' choice. Instead, it constitutes an institutional arrangement. According to Orsini (2014: 2), maritime arrivals after search and rescue (SAR) incidents are not actual arrivals. They are part of an institutional decision, as migrants, after being rescued, are then transferred and disembarked to a harbour selected by authorities. Indeed, Italian authorities declared Lampedusa an unsafe harbour, opting to disembark migrants to other destinations.

Irregular migration flows towards Lampedusa were further cut down in 2018 after the decision of the then right-wing Minister of Interior, Matteo Salvini, to block charity vessels from disembarking irregular border crossers in Italian ports. In 2021, in spite of restrictions on movement, like quarantine for newcomers, imposed on account of the COVID-19 pandemic, the newly appointed Mario Draghi's government faced a significant increase in migratory flows, which rendered Lampedusa once again the main landing point for irregular border crossers arriving in Italy. This trend seems to be carrying on in 2022, as the island is now recording increased arrivals, especially during the summer months.

74 *Constructing the border of Lampedusa*

Regarding migrants' sea route, the majority of irregular sea border crossers that arrive in Lampedusa are facilitated by smuggling networks (Frontex, 2018a: 35). Libya, Tunisia, and Algeria are the main departure countries (UNODC, 2018: 145). Yet, for many, this journey did not start at the shores of these countries. Instead, the departure point was their home in West or East Africa, the Middle East, or even Asian countries, such as Bangladesh, indicating a longer and tiring route.

Their sea journey usually starts during the evening to avoid interception before entering international waters and can last more than two days (UNHCR et al., 2017: 105). Though tariffs can be negotiated and vary according to nationalities, in general, each migrant pays more than $1,000 to be smuggled from Libya to Italy (RMMS, 2014: 48). This journey usually does not take place in the season of winter, to avoid extreme weather conditions in the open sea (Interviewee 12). The sea vessels used are often small wooden boats, even manufactured locally, or rubber inflatable dinghies (House of Lords, 2016: 13). Most are unseaworthy, without navigation tools, overcrowded, with insufficient fuel and food on board (Frontex, 2016: 46). For each journey, the wooden boats can be packed with more than 700 people on board, whereas the respective number for inflatable dinghies can reach 140 persons (UNHCR et al., 2017: 6). Yet, in 2020, smugglers altered their modus operandi by opting for better quality and less overcrowded vessels in order to avoid interception in North African waters (Frontex, 2021e: 16). In parallel, a new trend started to be observed of migrants organising and conducting this sea journey in small groups with their own sea vessels and without the facilitation of smuggling networks (Giliberti & Queirolo Palmas, 2021: 5). The sea vessels' unseaworthiness and frequent overcrowding, Lampedusa's rocky coastline that hinders disembarkation, the perilous sea, and the severe open-sea weather render it the deadliest migratory route in the world, leading to the massive death toll of 4,578 recorded only in 2016 (UNHCR, 2022).

All these have had an impact on Lampedusa as a border, island, and community. Actually, the island's cross-cultural heritage, its alienation from the Italian capital as well as shipwrecks' fatalities have constructed Lampedusa as a community of openness,[5] tolerance, and solidarity (Melotti et al., 2018). The island's inhabitants, especially fishermen, many times have rescued or tried to rescue irregular migrants in defiance of any legal actions against them, such as the confiscation of their boats (Puggioni, 2015: 1149). At the same time, Lampedusa's community has been repeatedly nominated for the Nobel Peace Prize, producing even a document for migrants' rights entitled 'Charter of Lampedusa' (Carta di Lampedusa). In the memory of all the border crossers that have lost their lives during this sea journey, a monument has been erected on this island that forms a door to Europe (Porta d'Europa), marking this link between Africa and Europe.

Hence, Lampedusa is characterised by pairs of dichotomies, like North/ South, (Continental Europe and Mediterranean), South/North (Southern

Europe and Northern Africa), and South/East (Southern Europe with Middle East) (Giuliani, 2018). All these dichotomies have framed this border in an antithetic context marked, at the same time, by its fluid environment as a remote island. On top, in the last decades it has evolved into a border spectacle (Cuttitta, 2012, 2014), due to the border control policies implemented by the actors involved in border control.

Border control conduct and actors

In Italy, border control is being entrusted upon different actors. This leads to a fragmented border control conduct. Generally, three authorities are responsible for the country's border control and border security: the Italian State Police (Polizia di Stato), the National Military Police Corps (Carabinieri), and the Finance Police (Guardia di Finanza). For sea borders, maritime patrolling is being shared among the Finance Police, the Italian Coastguard (Guardia Costiera), and the Navy (Marina Militare). The Navy acts in international waters, the Finance Police within the 24 nautical miles zone, while the Coastguard undertakes SAR operations.

The Navy is a military force tasked with the country's maritime defence and the combat of trafficking. The Finance Police is a special military police force under the control of the Ministry of Economy and Finance. Its tasks include the combat of smuggling and irregular migration. The Coastguard is a military force that is part of the Italian Navy. However, it is not oversighted by the Ministry of Defence. Instead, it operates under the auspices of the Ministry of Infrastructures and Transport. The Coastguard is responsible for the safety of navigation and the protection of human life at sea, monitoring and undertaking SAR operations not solely to the Italian territorial waters but to the 'entire region of interest on the Italian sea' (Guardia Costiera, 2022). Also, in ports, there is the Customs and Monopolies Agency (Agenzia delle Dogane e dei Monopoli—ADM) controlling goods entering or exiting the national maritime borders.

These actors are also present in Lampedusa. On the island there is a local station of the National Military Police, two offices of the Finance Police, an ADM office for customs, a Coastguard office, an aeronautical military base, and a remote radar military station integrated in NATO's air defence system. The numerical presence of uniformed staff is substantial. In 2009, after the Tunisian riots, there were 450 police and military officers, amounting to a ratio of 1:13 inhabitants (Cuttitta, 2014: 205). In general, most personnel stationed in Lampedusa are relatively young. Officers that are in their first years of their career are transferred there, usually for a limited-duration service. Although the harsh conditions in the sea require the high preparedness and vigilance of the younger personnel, the unpredictable sea environment coupled with human losses at sea due to the frequent shipwrecks, from a psychological viewpoint, may have a detrimental impact on younger and less experienced officers.

The technical means used for border control are predominantly maritime and air, coupled with surveillance tools. The naval assets employed include the Coastguard's motorboats and coastal patrol boats as well as National Military Police motor patrol vessels. Maritime surveillance is being conducted with helicopters, aeronautical systems, satellites, and radar systems. Actually, a new long-range radar system has been installed, expanding the surveillance area up to 470 km (Difesa, 2019). This means that maritime patrolling exceeds the Italian territorial waters.

After all, in SAR cases, SAR and then interception take place before the migrants' boat reaches Lampedusa's shore. SAR operations begin with the receipt of information about a migrant boat in danger of sinking. After the localisation of the boat's position using naval and air assets, a SAR team intervenes by approaching the boat in distress. Next, migrants are transferred usually to Coastguard motorboats in order to disembark safely in a designated pier, that is, the military and gated quay of Favarolo, opposite to Lampedusa's old port. After their rescue and the provision of medical assistance, irregular migrants are then transferred to the hotspot facility so as to be identified, registered, and fingerprinted. Yet, in the aftermath of COVID-19, a sea-quarantine on ships' procedure was applied to newcomers in an attempt to control movement and virus spread (Giliberti & Queirolo Palmas, 2021).

Frontex on the field

Italian authorities are not the sole actor present at the island. Ever since 2006, when Frontex coordinated its joint operation 'Nautilus', the agency has developed a continuous presence at the sea border of Lampedusa, aiding Italy in the management of irregular migration flows. Frontex's role entails assistance in sea border control, border surveillance, the fight against irregular migration and cross-border crime, as well as SAR activities in the Central Mediterranean (Interviewee 10).

Regarding joint operations, besides Nautilus, Frontex has coordinated the joint operation 'Hermes' (initiated in 2007), the 'EPN-Aeneas' (from 2011 until 2014), the 'EPN-Triton' (from 2014 until 2018), while, since 2018, it has conducted the 'Themis' operation. To fulfil Themis' mandate, Frontex, from the very beginning, has deployed 260 officers and diverse technical equipment, including naval vessels, fixed-wing aircrafts, and helicopters for aerial surveillance (European Commission, 2018).

Apart from joint operations, Frontex also supports Italy in the implementation of the hotspot approach. Doing so, Frontex deploys support officers and experts at the Italian hotspots, such as Lampedusa, to assist Italian authorities with the registration, screening, and processing of irregular migrants. Approximately 25 Frontex officers are in each Italian hotspot facility (Frontex, 2017c: 11).

Furthermore, Frontex set up in Catania, in Sicily, an EU Regional Task Force (EURTF) office, which provides logistical and operational support to Frontex's operations in Italy. This office also cooperates with Frontex's liaison officer stationed in Rome. Moreover, in 2021, enhancing its role in returns, Frontex deployed its first return team, consisting of 11 officers from its standing corps, to be based at Rome's Fiumicino airport. Additionally, Frontex has conducted various training curricula in the Italian peninsula, like survival in cold weather conditions as well as law enforcement and SAR.

Beyond Frontex's activity in the Italian waters, the agency also promotes its role on the other side of the border. Although its Working Arrangements with Libya and Tunisia have not yet been finalised, Frontex found a way to bypass this omission in order to cooperate with these countries. Regarding Libya, since 2006 it has started negotiating the signature of a Working Arrangement. Despite its non-finalisation, in 2007 it led a technical mission on irregular migration and border control to Libya. This mission included field trips, meetings with Libyan authorities as well as information exchange, laying the foundation for an operational cooperation between Libya and Frontex. Following that, Libyan representatives participate in Frontex meetings, while Frontex trains Libyan coastguard and navy officers (Frontex, 2019a). In parallel, Frontex cooperates with the EU Border Assistance Mission (EUBAM) Libya,[6] deploying Frontex experts that provide their border management expertise. Also, in 2018, Frontex opened its first Risk Analysis Cell in Niger, which borders Libya. This Cell is run by local analysts trained by Frontex with the mission to collect and analyse strategic data on migration and border management. Prioritising North Africa, the same year Frontex's management board decided to deploy a Frontex Liaison Officer (FLO) in Tunisia, though this deployment — at the time of writing these pages — is yet to take place. Moreover, deepening its cooperation with African countries, it has developed an Africa-Frontex Intelligence Community (AFIC) for intelligence sharing and joint risk analysis. In AFIC, Libya is a full participant, whereas Tunisia has observer status.

The above show that since 2006 Frontex has developed a continuous presence at the border of Lampedusa, impacting diversely the border control conduct (see Table 4.2). It coordinates and implements sea operations, it brings technical and human resources, like the 440 officers deployed during operation Triton, it engages with hotspots, it enables Italian officers to gain experience and adapt their work methods by participating in sea, land, and air operations coordinated by Frontex outside Italy, like Hera, Indalo, and Poseidon Land, as well as it trains and educates Italian officers, such as through the Maritime Border Surveillance Officer training course that aims at establishing common standards for maritime border control, surveillance, and SAR. Moreover, it consults with the Italian border authorities, hosting frequent meetings at its headquarters in Warsaw. In parallel, it has found a way to cooperate with North African countries neighbouring Italy, namely Libya and Tunisia. According to Frontex, this cooperation has contributed to the decrease of flows in Central Mediterranean. The 'intensive

Table 4.2 Frontex in Lampedusa

	Activity	Scope	Composition
Sea Operations	Nautilus	surveillance, migration management, operational coordination	experts
	Hermes	surveillance, migration management, cross-border crime	experts (incl. debriefers, screeners)
	Aeneas	surveillance, migration management, cross-border crime	experts (incl. debriefers)
	Triton	SAR	SAR units, screeners, debriefers
	Themis	law enforcement, surveillance, SAR, migration management	experts (incl. debriefers)
Other	Cooperation with CSDP (Sophia, EUBAM)		
	Training		
	Hotspots		
	Returns		
	EURTF	regional support, situational awareness, operational management	6-person staff
	External relations with Libya & Tunisia	technical projects, technical missions, training, EUBAM Libya	FLO, AFIC

patrolling' by Libyan officers led to an 81% decrease in irregular migrant departures from Libya during the first quarter of 2018 compared to 2017 (Frontex, 2018b: 11). As a result, Frontex's role in Lampedusa is that of key border control actor that shapes the border control conduct.

Border control assumptions and practices in Lampedusa

The previous description of the border control conduct in Lampedusa reveals a pattern and a habitus for border control (Bourdieu, 1977). This pattern reflects cultural traits composed of background assumptions and practices for the border control conduct. Background assumptions include the implemented border control approach and border perceptions. Practices correspond to what border control officers commonly do when engaging with border control, and how they do it (Zaiotti, 2011: 26).

Starting with border assumptions, Lampedusa is both a national and an EU external border. On the one hand, the EU dimension validates Frontex's presence at this border, enabling at the same time border control officers from other EU countries to assist Italy in border control. Illustrative of that is the following assessment about Frontex's role at the border:

> [Border official] 'Frontex offers us assistance'
>
> (Interviewee 10).

On the other hand, the national context accounts for Lampedusa being militarised. Military personnel are stationed at Lampedusa, whereas across the island, there are more than 10 military zone signs. Yet, this military context does not stem from any quarrel with neighbouring countries or fear for Italy's territorial integrity. Instead, it emanates from the border control approach adopted at Lampedusa that advances a militaristic framework.

Regarding border control assumptions, Lampedusa border, as referred to above, unfolds a militarisation. The Italian authorities involved with maritime border control, namely the navy and the coastguard, belong to the military corps. Similarly, the Finance Police is a military police force. This means that their organisation, structure, means, operational concept, and doctrines are framed within a militaristic spirit (Jones & Johnson, 2016). So, they perceive and refer to border control with military terms, as attested discursively:

> [Border official] 'We monitor the target'
>
> (Interviewee 10).

That can also be extracted from the selected wording in official publications about migration, for example, '[...] activation performed by air-sea military and civil assets, sighting units in danger' (Guardia Costiera, 2017: 20). In turn, irregular migration is being managed with military operations, such as the Italian military-humanitarian operation 'Mare Nostrum' conducted in 2013 in the Sicilian Strait. To manage migration in the Central Mediterranean, Frontex also started to cooperate with CSDP's naval operation 'Sophia' by exchanging liaison officers and sharing information on the location of vessels detected and the position of naval assets (Frontex, 2017b: 28). Hence, Frontex, which constitutes a border agency that deals with non-traditional security challenges, established a channel of cooperation with a military operation, despite its different mandate, scope, means, and staff.

80 *Constructing the border of Lampedusa*

Militarisation is accompanied by securitisation. Securitisation is characterised by conditions of exception and emergency (Buzan et al., 1998: 21–26). Lampedusa has been repeatedly declared in a state of emergency due to migration flows. For instance, in 2011, extraordinary powers were given to regional authorities to manage the mass migratory flow. In the same context, Rome requested from Frontex to reinforce its Hermes operation due to the 'extraordinary migratory situation' (Frontex, 2011). The emergency state was coupled with emergency policies, measures as well as an emergency discourse denoting Lampedusa's securitisation (Campesi, 2011). For instance, official publications on migration both show and reproduce a securitisation context by invoking a state of 'ongoing critical emergency' and describing an 'emergency situation [that] keeps being severe' (Guardia Costiera, 2017). Even the current Frontex operation, Themis, has an important security component that involves the detection of terrorist threats at the external border (Frontex, 2022), manifesting, as a result, a security focus.

Another border control assumption is technocracy. A managerial and technocratic border control approach is encountered at Lampedusa (Cuttitta, 2018: 636). Technocracy in Lampedusa's border control involves the deployment of 'specifically adapted assets and specially trained crews', namely 'authentic [...] maritime experts' (Guardia Costiera, 2022). Accordingly, border control is being conducted with the use of computer systems and experts, such as debriefing and screening experts. Geospatial imagery is also used, such as with the application of EUROSUR, which enables border control officers to track possible boat arrivals and monitor them throughout their sea route.

This long-distance tracking manifests that border control does not take place at the space of the border. Actually, no concrete line can demarcate a sea border. The sovereignty of a state extends to the sea. As a result, both the territorial waters and the contiguous zone correspond to state rights and control. In parallel, they include obligations that emanate from the Law of the Sea (UNCLOS) and the protection of persons in distress at sea. For this reason, border control and sea patrolling has been extended not only to the territorial waters of the countries neighbouring Lampedusa, but also across the whole Mediterranean Sea region. Reflecting that, according to the Italian authorities, the 'region of interest on the Italian sea [...] overcomes the territorial waters boundaries' (Guardia Costiera, 2022). The extension of border control is being implemented with the installation and use of long-range radars. This reflects an extra-territorialisation of both border and border control. An example of this extra-territorialisation constitutes operation Triton. According to Triton's operational plan, all irregular migrants intercepted or rescued were to be disembarked in Italy (Frontex, 2015: 6). Yet, most SAR incidents were taking place close to the Libyan shore, which was clearly outside Frontex's operational area. Given that Libyan authorities did not have the capacity to act, Frontex, responding to SAR emergencies, started operating outside its members' territory, conducting out-of-area border patrolling and long-range surveillance. To do so, it deployed maritime assets enabling offshore patrolling in the open sea, like offshore patrol vessels used in Hermes and Triton Frontex operations. Nevertheless, the operational area has been a matter of juxtaposition, as it was considered a pull factor attracting irregular migrants (Frontex, 2017c: 3). In this context, Italian authorities pushed for a downsized operational area in Frontex operations, as with Themis. Themis' patrolling area is up to 24 nautical miles from the Italian shore. Hence, it does not cover SAR activities on the Libyan coast. This mirrors a parallel intra-territorialisation limiting the responsibility area for migration and therefore the number of disembarked irregular migrants in the Italian territory.

Border control in Lampedusa is also predominated by surveillance; in other words, border control emphasising on borders' monitoring (Interviewee 10).

[Border official] 'The most important thing is surveillance'
(Interviewee 10).

Surveillance at sea borders permits not only to monitor the border but also to rescue more persons in distress at sea (Jumbert, 2018). For this purpose, radars, sensors, and satellite monitoring systems have been installed in Lampedusa to track vessels, like the Vessel Monitoring System (VMS) and the Long-Range Identification and Tracking (LRIT). This surveillance context is also reflected in Frontex's operations. For instance, during the 2008 Nautilus operation, Frontex started using maritime surveillance services including earth observation with satellite images for sea vessel monitoring. Enhancing aerial surveillance, in 2015, Frontex developed a new concept that it employed during operation Triton. This concept involved the deployment of fixed-wing aircrafts that stream aerial imagery data directly from the border to Frontex Situation Centre at Frontex's headquarters. Frontex's surveillance facilitates the detection of small boats that have high speed, which constituted 'a great problem' for the Italian authorities and 'was ensured by Frontex' (Guardia Costiera, 2017: 30).

Intelligence constitutes another border control characteristic encountered in Lampedusa. Almost every Frontex operation in Italy lists intelligence as an operational priority. Following this spirit, currently, Themis operation emphasises the collection of intelligence in order to detect terrorists at the border (Frontex, 2022). Intelligence gathering is implemented especially through debriefing. Actually, in 2016, during operation Triton, Frontex guest officers deployed as debriefers interviewed 2,400 irregular migrants (Frontex, 2017a: 36). Besides debriefing, intelligence is also collected via risk analysis fostered through the AFIC network. In fact, AFIC was set up having as goal to improve data collection, sharing, and analysis by building a framework for regular information-sharing on border-related matters between the EU and African countries (Frontex, 2017d).

Regarding border control practices, Lampedusa's border control is implemented via SAR. The majority of sea arrivals to Lampedusa involve SAR activities (Interviewee 10). In 2017, from the 119,445 irregular sea migrants that arrived in Italy, 114,286 were rescued during SAR operations (Guardia Costiera, 2017). SAR is also included as an objective in each Frontex sea operation. Moreover, the SAR area, the competent SAR authority, the rules of engagement during SAR operations, as well as the disembarkation port after SAR incidents, are all part of a particular operational methodology applied to border control during SAR (Regulation, 2014). Aside from border control, SAR's predominance affects also the border control perceptions. Ship wreckages, deaths, and persons in distress at sea have become synonymous with border control in Lampedusa, affecting policies and public attitudes. Therefore, SAR is simultaneously a border control driver and a corollary (Carrera & den Hertog, 2015: 3).

The border of Lampedusa functions also as an information gathering hub. Data like geospatial information on boats' sea routes and intelligence collected during debriefing contribute to border control, rendering, in turn, Lampedusa an information gathering and sharing environment. This is facilitated through Frontex, which has developed common platforms, like JORA, so that border control officers can upload operational data on border control incidents. Likewise, the EURTF in Catania smooths information exchange by gathering data and then channelling them to the relevant actors.

From the above, it is possible to discern that border control in Lampedusa is implemented based on multilateral cooperation. Guest officers from various EU countries as well as Frontex coastguards from Frontex's newly-formed standing corps arrive in Lampedusa to participate in Frontex joint operations. Currently (June 2022), 248 officers from Frontex's standing corps are being deployed in Italy to assist their Italian colleagues. This reflects a community of border control practitioners built upon a common goal, that is the protection of this EU external border. In addition, Italian authorities collaborate and communicate daily with Frontex officers (Interviewee 13). This cooperation is being also spread to the other side of the border. Italian authorities and Frontex cooperate with Libya and Tunisia in the field of migration. In this context, mixed crews comprising Italian and Libyan border control officers have patrolled the Mediterranean Sea, while in 2017 Italy and Libya deepened their cooperation by signing a Memorandum of Understanding, which led Italian authorities to donate patrol boats to Libya and train Libyan officers. Besides, Lampedusa has become a symbolic meeting place for migration and border control discussions fostering dialogue and cooperation among EU countries (Ministero dell'Interno, 2017).

The use of technology constitutes another practice for border control. Border control in Lampedusa relies on a sophisticated electronic system, which encompasses radars, satellites, and helicopters. Many tools of this system have been developed in the context of EU-funded security research projects, like the EUCISE2020. The centrality of technology is also manifested with the implementation of EUROSUR, which uses state-of-the-art satellite radar technology, such as automated vessel tracking and position prediction software, including even data from Copernicus space system. Recently, after trials in Lampedusa, Frontex started deploying drones in its operations at the island (Interviewee 11; Frontex, 2019a). All these describe a sophisticated technological border control.

Border control in Lampedusa is also professionalised. According to the Italian authorities, sea border control has 'entered into a highly professional stage' using specialised human and technical resources (Guardia Costiera, 2022). Experts, namely professionals with extensive skills and experience, having variant profiles, such as screening experts, debriefing experts, and SAR experts, are deployed to Lampedusa, especially during Frontex missions, to provide their professional expertise. They cooperate with officers from different backgrounds, like military, police, and customs, developing an operational policy community of professionals (Carrera & den Hertog, 2015: 20). They also work with colleagues transferred to Lampedusa from other Italian cities. For instance, during the Italian Mare Nostrum operation, more than 700 officers were seconded to the island. Moreover, border control officers' work is being periodically assessed in terms of shifts in migratory flows and death incidents. In fact, Frontex publishes sea surveillance reports and develops rules, work standards, and training curricula for maritime surveillance. All these advance border control's professionalisation in Lampedusa.

These underlying assumptions and practices represent a routinised mode for border control, implemented and reproduced at the border by the border control practitioners (see Table 4.3). As such, they are cultural traits forming the border control conduct in the border of Lampedusa. In parallel, they

Table 4.3 Border control cultural traits in Lampedusa

Border assumptions	EU external border – post-national integration
	National territory – militarisation
Border control assumptions	Militarisation
	Securitisation
	Extra-territorialisation/Intra-territorialisation
	Technocracy
	Surveillance
	Intelligence
Border control practices	SAR
	Information gathering
	Multilateral cooperation
	Technology
	Professionalisation

construct the border itself by being integrated in the implemented approach and shared vision for Lampedusa's border management.

Conclusion

After the analysis on Frontex (Chapters 2 and 3) and keeping in mind that 'topos' matters, this chapter turned to the location of border control, to the border itself, studying the Italian sea border of Lampedusa. In the last decades, Lampedusa has become a symbolic EU border and a 'theatre of the border play' (Cuttitta, 2014), partly due to the vast irregular migration that keeps attracting and partly as a result of the border control policies adopted to manage migration. Yet, being an EU external border, which hosts numerous Frontex operations, it allows researching both border control and Frontex's role in border control.

The chapter, first, referred to Lampedusa's geographical characteristics, addressed its border development, and looked into its border control function. Then it analysed the border control approach implemented there, extracting cultural traits, namely border control assumptions and practices that characterise the border control conduct at this border. Lampedusa's analysis revealed that this border has been transformed. Lampedusa's transformation stems from its dual attribute that renders Lampedusa as both a national and an EU external border. Furthermore, the chapter attested Frontex's role as a border control actor that acts at the border, interacts with the other border control practitioners, and conducts border control. Actually, Frontex through its function, advances and promotes some of the assumptions and practices traced at the border. These assumptions and practices both construct and govern Lampedusa as a border and 'topos' of border control.

But are these signals for a new border control culture? To answer this question, the analysis moves to another EU external border, that is, the Greek land border of Evros, carrying on the comparative case study analysis with a different type of border.

Notes

1 The Central Mediterranean sea route refers to migratory flows from North Africa to Italy and Malta.
2 Push-back practices refer to the interception of irregular border crossers on the high seas and then their return to Libya. These practices breach the principle of non-refoulement. For refoulement, see Papastavridis (2010).
3 On 17 December 2010, a young Tunisian set himself on fire to protest against the Tunisian authoritarian regime of President Zine el-Abedin Ben Ali. The subsequent street protests that broke out across the country, besides the regime's violently response, led to the overthrow of the Tunisian President. The so-called 'Jasmine Revolution' in Tunisia ignited as a chain reaction the Arab Spring uprisings in the countries of North Africa and Middle East, such as Libya, Egypt, Syria, and Yemen (Khan & Mezran, 2015).

4 The EU relocation scheme was a programme that covered the transfer of asylum seekers from Italy and Greece to other European states. It applied only to applicants for which the average recognition rate of international protection at the EU level was above 75%. The programme ended on 26 September 2017.
5 However, according to Orsini's analysis (2015), this pro-migrant attitude is not holistic especially towards Tunisians.
6 EUBAM Libya is a Common Security and Defence Policy (CSDP) mission launched in 2013.

References

Bernardie-Tahir, N. & Schmoll, C. (2014) 'Islands and Undesirables: Introduction to the Special Issue on Irregular Migration in Southern European Islands'. *Journal of Immigrant & Refugee Studies*, 12(2), pp. 87–102.
Bourdieu, P. (1977) *Outline of a Theory of Practice*. Cambridge: Cambridge University Press.
Brambilla, C. (2016) 'Navigating the Euro/African Border and Migration Nexus through the Borderscapes Lens: Insights from the LampedusaInFestival'. In: C. Brambilla, J. Laine, J.W. Scott & G. Bocchi (eds.), *Borderscaping: Imaginations and Practices of Border Making*. Abingdon: Routledge, pp. 111–121.
Buzan, B., Wæver, O. & de Wilde, J. (1998) *Security: A New Framework for Analysis*. Boulder: Lynne Rienner.
Campesi, G. (2011) 'The Arab Spring and the Crisis of the European Border Regime: Manufacturing Emergency in the Lampedusa Crisis'. *EUI*, Working Paper 2011/59, pp. 1–21.
Carrera, S. & den Hertog, L. (2015) 'Whose Mare? Rule of Law Challenges in the Field of European Border Surveillance in the Mediterranean'. *CEPS*, Paper in Liberty and Security in Europe 79, pp. 1–29.
CoE (2011) 'Report on the Visit to Lampedusa (Italy)'. *Committee on Migration, Refugees and Population*. Strasbourg: Parliamentary Assembly.
Cresti, F. (2008) 'Comunità Proletarie Italiane nell'Africa Mediterranea tra XIX Secolo e Periodo Fascista'. *Mediterranea*, V(12), pp. 189–214.
Cuttitta, P. (2012) *Lo Spettacolo del Confine: Lampedusa tra Produzione e Messa in Scena della Frontiera*. Milan: Mimesis.
Cuttitta, P. (2014) 'Borderizing the Island Setting and Narratives of the Lampedusa Border Play'. *ACME: An International E-Journal for Critical Geographies*, 13(2), pp. 196–219.
Cuttitta, P. (2018) 'Repoliticization through Search and Rescue? Humanitarian NGOs and Migration Management in the Central Mediterranean'. *Geopolitics*, 23(3), pp. 632–660.
Difesa (2019) 'New Radar for Surveillance of National Airspace Inaugurated in Lampedusa'. Available at http://en.difesaonline.it/news-forze-armate/cielo/inaugurato-lampedusa-nuovo-radar-la-sorveglianza-dello-spazio-aereo (accessed February 2022).
ECtHR (2012) Case of Hirsi Jamaa and Others v. Italy. App. No. 27765/09.
European Commission (2013) 'Statement by President Barroso Following his Visit to Lampedusa'. Available at https://ec.europa.eu/commission/presscorner/detail/fr/SPEECH_13_792 (accessed January 2022).

European Commission (2018) 'Managing Migration in all its Aspects: Commission Note'. Available at https://op.europa.eu/en/publication-detail/-/publication/4355a9f7-ed3f-11e8-b690-01aa75ed71a1 (accessed January 2022).

Frontex (2011) 'Hermes 2011 Starts Tomorrow in Lampedusa'. Available at https://frontex.europa.eu/media-centre/news/news-release/hermes-2011-starts-tomorrow-in-lampedusa-X4XZcr (accessed February 2022).

Frontex (2015) *Frontex Annual Report on the Implementation on the EU Regulation 656/2014*. Warsaw: Frontex.

Frontex (2016) *Risk Analysis for 2016*. Warsaw: Risk Analysis Unit.

Frontex (2017a) *Risk Analysis for 2017*. Warsaw: Risk Analysis Unit.

Frontex (2017b) *Annual Activity Report 2016*. Warsaw: Frontex.

Frontex (2017c) *JO Triton 2017: FRONTEX Evaluation Report*. Warsaw: Frontex.

Frontex (2017d) 'Frontex Launches Capacity Building Project for Africa during AFIC Meeting'. Available at https://frontex.europa.eu/media-centre/news/news-release/frontex-launches-capacity-building-project-for-africa-during-afic-meeting-g3JfQJ (accessed February 2022).

Frontex (2018a) *Risk Analysis for 2018*. Warsaw: Risk Analysis Unit.

Frontex (2018b) *FRAN Quarterly: January-March 2018*. Warsaw: Risk Analysis Unit.

Frontex (2019a) 'Law Enforcement and SAR Training Completed in Ostia'. Available at https://frontex.europa.eu/media-centre/news/news-release/law-enforcement-and-sar-training-completed-in-ostia-JAK7yu (accessed February 2022).

Frontex (2019b) 'Frontex Detects Mother Boat Smuggling People'. Available at https://frontex.europa.eu/media-centre/news/news-release/frontex-detects-mother-boat-smuggling-people-dIBt9Q (accessed February 2022).

Frontex (2021) *Risk Analysis for 2021*. Warsaw: Risk Analysis Unit.

Frontex (2022) 'Operation Themis'. Available at https://frontex.europa.eu/we-support/main-operations/operation-themis-italy-/ (accessed February 2022).

Giliberti, L. & Queirolo Palmas, L. (2021) 'The Hole, the Corridor and the Landings: Reframing Lampedusa through the COVID-19 Emergency'. *Ethnic and Racial Studies*. https://doi.org/10.1080/01419870.2021.1953558 (published online).

Giuliani, G. (2018) 'The Colour(s) of Lampedusa'. In: G. Proglio & L. Odasso (eds.), *Border Lampedusa: Subjectivity, Visibility and Memories in Stories of Sea and Land*. London: Palgrave Macmillan, pp. 67–85.

Global Initiative (2014) 'Smuggled Futures: The Dangerous Path of the Migrant from Africa to Europe'. Research Report May 2014.

Guardia Costiera (2017) 'SAR Operations in the Mediterranean Sea'. *Maritime Rescue Coordination Centre Roma*, Report, pp. 1–32.

Guardia Costiera (2022) 'Search and Rescue'. Available at https://www.guardiacostiera.gov.it/en/Pages/search-and-rescue.aspx (accessed March 2022).

House of Lords (2016) 'Operation Sophia, the EU's Naval Mission in the Mediterranean: An Impossible Challenge'. *European Union Committee*, HL Paper 144.

IOM (2018) 'Mixed Migration Flows in the Mediterranean: Compilation of Available Data and Information'. Report March 2018.

Jones, R. & Johnson, C. (2016) 'Border Militarisation and the Re-Articulation of Sovereignty'. *Transactions of the Institute of British Geographers*, 41(2), pp. 187–200.

Jumbert, M.G. (2018) 'Control or Rescue at Sea? Aims and Limits of Border Surveillance Technologies in the Mediterranean Sea'. *Disasters*, 42(4), pp. 674–696.

Khan, M. & Mezran, K. (2015) 'Tunisia: The Last Arab Spring Country'. *Atlantic Council*, Issue Brief, pp. 1–10.

Kitagawa, S. (2011) 'Geographies of Migration across and Beyond Europe: The Camp and the Road of Movements'. In: L. Bialasiewicz (ed.), *Europe in the World: EU Geopolitics and the Making of European Space*. Farnham: Ashgate, pp. 201–222.

Li Causi, L. (1987) 'Lampedusa: The Road to Marginality'. *Ekistics*, 54(323/324), pp. 165–169.

Melotti, M., Ruspini, E. & Marra, E. (2018) 'Migration, Tourism and Peace: Lampedusa as a Social Laboratory'. *Anatolia*, 29(2), pp. 215–224.

Ministero dell'Interno (2017) 'Sicurezza nell'Area del Mediterraneo: Riunione dei Capi della Polizia'. Available in Italian at http://www.interno.gov.it/it/galleria-video/sicurezza-nellarea-mediterraneo-riunione-dei-capi-polizia (accessed March 2022).

Ministero dell'Interno (2022) 'Cruscotto Statistico Giornaliero'. Available in Italian at http://www.libertaciviliimmigrazione.dlci.interno.gov.it/it/documentazione/statistica/cruscotto-statistico-giornaliero (accessed March 2022).

Newman, D. (2006) 'Borders and Bordering: Towards an Interdisciplinary Dialogue'. *European Journal of Social Theory*, 9(2), pp. 171–186.

Newman, D. (2011) 'Contemporary Research Agendas in Border Studies: An Overview'. In: D. Wastl-Water (ed.), *The Ashgate Research Companion to Border Studies*. Farnham: Ashgate, pp. 33–47.

Orsini, G. (2014) 'Testing Societal Security at the Border: The Case of Lampedusa'. *Source*, September, pp. 1–3.

Orsini, G. (2015) 'Lampedusa: From a Fishing Island in the Middle of the Mediterranean to a Tourist Destination in the Middle of Europe's External Border'. *Italian Studies*, 70(4), pp. 521–536.

Papastavridis, E. (2010) 'Fortress Europe and FRONTEX: Within or Without International Law?'. *Nordic Journal of International Law*, 79(1), pp. 75–111.

Popescu, G. (2012) *Bordering and Ordering in the Twenty-First Century: Understanding Borders*. Lanham: Rowman and Littlefield.

Puggioni, R. (2015) 'Border Politics, Right to Life and Acts of Dissensus: Voices from the Lampedusa Borderland'. *Third World Quarterly*, 36(6), pp. 1145–1159.

Regulation (EU) No 656/2014 of the European Parliament and of the Council of 15 May 2014 Establishing Rules for the Surveillance of the External Sea Borders in the Context of Operational Cooperation Coordinated by the European Agency for the Management of Operational Cooperation at the External Borders of the Member States of the European Union [2014, OJ L 189/93].

RMMS (2014) 'Going West: Contemporary Mixed Migration Trends from the Horn of Africa to Libya & Europe'. Study 5.

Shaw, M. & Mangan, F. (2014) 'Illicit Trafficking and Libya's Transition: Profits and Losses'. *US Institute of Peace*, Peaceworks No. 96.

Tazzioli, M. (2018) 'Containment through Mobility: Migrants' Spatial Disobediences and the Reshaping of Control through the Hotspot System'. *Journal of Ethnic and Migration Studies*, 44(16), pp. 2764–2779.

UNHCR (2022) 'Mediterranean Situation: Italy'. Available at https://data2.unhcr.org/en/situations/mediterranean/location/5205 (accessed January 2022).

UNHCR, IMPACT & Altai Consulting (2017) 'Mixed Migration Trends in Libya: Changing Dynamics and Protection Challenges'. Available at http://www.unhcr.org/595a02b44 (accessed March 2022).

UNODC (2018) 'Global Study on Smuggling of Migrants: Europe'. UNODC Research.

van Houtum, H. & Mamadouh, V. (2008) 'The Geopolitical Fabric of the Border Regime in the EU-African Borderlands'. In: V. Mamadouh, S.M. de Jong, F. Thissen, J. van der Schee & M. van Meeteren (eds.), *Dutch Windows on the Mediterranean: Dutch Geography 2004-2008*. Utrecht: KNAG, pp. 93–99.

van Houtum, H., Kramsch, O. & Ziefhofer, W. (eds.) (2005) *B/ordering Space*. Aldershot: Ashgate.

WHO (2012) 'Increased Influx of Migrants in Lampedusa, Italy'. Joint Report.

Wilson, T.M. & Donnan, H. (1998) 'Nation, State and Identity at International Borders'. In: T.M. Wilson & H. Donnan (eds.), *Border Identities: Nation and State at International Frontiers*. Cambridge: Cambridge University Press, pp. 1–30.

Zaiotti, R. (2011) *Cultures of Border Control: Schengen & the Evolution of European Frontiers*. Chicago: University of Chicago Press.

Interviews

Interviewee 10: Border official in Lampedusa, 10 May 2018.
Interviewee 11: Border officer in Lampedusa, 10 May 2018.
Interviewee 12: Border officer in Lampedusa, 11 May 2018.
Interviewee 13: Border officer in Lampedusa, 12 May 2018.

5 Constructing the border of Evros

'[Evros] border is not only a Greek border
but it is also a European border [...]
The situation at our border is not only Greece's issue to manage.
It is the responsibility of Europe as a whole'.
—Ursula von der Leyen, current European Commission President
(European Commission, 2020)

Introduction

Each border is different. By being made (van Houtum et al., 2005; Newman, 2011) or, put differently, being constructed, each border is evolving (Newman, 2003) and shaped by the environment and the actors that act, are present, and operate at it. So, looking into just one border is simply not enough for reaching a broader verdict about border control at Europe's borders or Frontex's role in it. After all, the EU external border extends to 51,000 km, surpassing 800 border crossing points at both land and sea (Frontex, 2022a). At the same time, studying each and every one of these borders seems like an insurmountable task for just one person and way out of reach for the pages of just one book. To remedy this methodological challenge, I opted for variation. Variation allows drawing a wider conclusion about common patterns or diverging characteristics observed at different borders (Mill, 1865 [1843]). Keeping this in mind, I have chosen to compare two different EU external borders: Lampedusa and Evros. So, after analysing the sea border of Lampedusa in the previous chapter (Chapter 4), I will now delve into the land border of Evros (Chapter 5).

Lampedusa and Evros constitute two symbolic EU external borders, which have repeatedly become Frontex's operational focus due to the vast irregular migration flows that they keep attracting. Yet, they are completely different in geopolitical, symbolic, functional, and geographic terms. By comparing Lampedusa and Evros, this study can draw a wider conclusion about border control and Frontex's function at the EU external borders. This comparison can also produce new knowledge about borders and

DOI: 10.4324/9781003230250-5

border control in Europe, as Lampedusa and Evros have not been analysed comparatively before. So, Lampedusa and Evros is a pair that not only can but also deserves to be compared and analysed, given that irrespective of the outcome, their comparative analysis contributes original knowledge about EU borders and border control.

Against this background, the chapter moves the analysis to the Greek-Turkish land border of Evros. Continuing the comparative case study analysis, which started in the previous chapter with Lampedusa (Chapter 4), this chapter has the same goal, that is, to explore the border control conduct and border control actors that form this EU external border so as to extract the cultural traits that construct it. These traits are indicators of a border control culture at Europe's borders.

Maintaining a similar structure to Chapter 4 for the purposes of comparative case study analysis, the chapter proceeds with an exploration of the border setting. Then, it refers to the border control conduct and the actors involved in border control, including Frontex. The next part elaborates on the cultural traits encountered at the border, which relate to certain border control assumptions and practices. Lastly, it elaborates the findings drawn from the research of Evros and Lampedusa. The analysis for Evros border, like Lampedusa's, relies on data derived from in situ analysis with fieldwork in Evros and interviews with Greek officers.

This chapter argues that Evros has been transformed as a border. Its transformation is due to its EU external border status acquisition that now adds to Evros' national border role and can be demonstrated by the border control assumptions and practices that shape it as a border. In addition, the chapter empirically attests Frontex's border control actorness at the Evros border, derived from the agency's involvement with border management. On the comparative case study, it shows that Evros and Lampedusa, despite their inherent differences as borders, share certain common assumptions and practices.

The river setting

'Evros' in Greek language, 'Meriç' in Turkish, and 'Марица' in Bulgarian is the name of a river that flows through Greece, Turkey, and Bulgaria, forming a natural border among them. It constitutes the second largest river in the Balkans after the Danube river. Its total length is 530 km. This book's focus is solely the Greek-Turkish border, which sets the land frontier of the EU with Turkey.[1] This frontier extends for more than 200 km. Most of the length of the border is part of the river, except for 12.5 km of land strip.

Regarding the river features, like most rivers, Evros water is muddy, yet with a fast and sturdy current (Topak, 2014: 825). Its flow has gradually formed various lakes, lagoons, salt marshes, and low-tide elevations. On its banks there is significant vegetation including poplar, willow, elm trees, and bushes (Christianou, 2009). During winter, the temperature can drop well below zero degrees Celsius. The cold is harsh, often accompanied by humidity and fog that reduce visibility. There are also frequent incidents of heavy rainfalls that result in the river flooding.

The river has given its name to Evros prefecture in the Region of Eastern Macedonia and Thrace. Evros is the northernmost Greek prefecture, bordering with Turkey to the east and Bulgaria to the northwest. Evros prefecture covers an area of 4,242 km² and is composed of more than 130 villages and small towns as well as 146,582 inhabitants (Hellenic Statistical Authority, 2021). Evros prefecture is characterised by lowland areas, mostly of cotton cultivation, without mountainous territories or high hills. The prefectural capital, Alexandroupolis, has a harbour, a railway connection, and an airport with flights from and to Athens. Alexandroupolis is located 840 km from the Greek capital Athens, 295 km from the Turkish city Istanbul, and 153 km from the Turkish city Edirne. Even closer to Turkey is Orestiada, the second most populated town after Alexandroupolis, which is located in the northern region of Evros (see Map 5.1).

Evros' geographic remoteness from Athens and respective proximity to Turkey urge both Greece and Turkey to maintain a strong military presence in the region. This increased military deployment is due to the deep-rooted rivalry between Greece and Turkey. Notwithstanding their mutual NATO membership, the two neighbouring countries are traditionally considered adversaries. Actually, their rivalry is one of the oldest between neighbouring countries, bringing Greece and Turkey at various times to the verge of armed conflict (Moustakis & Sheehan, 2000: 95; Heraclides, 2012: 115). Since the fall of Constantinople in 1453 and Greece's Ottoman occupation that lasted from 1458 until the outbreak of the Greek War of Independence in 1821, Greeks have considered Turks a threat (Millas, 2004: 53; Theodossopoulos, 2006), or put differently, their 'stereotypical religious, ethnic, and cultural Other' (Argyrou, 2006: 33). Similarly, Turks perceive Greece as a hostile neighbour that may even initiate a military attack against Turkey (Çarkoğlu & Kirişci, 2004: 147). In this Greek-Turkish perpetual state of rivalry, critical was the Greek-Turkish War of 1919–1922 that led to the 'Asia Minor Catastrophe' and the Turkish invasion of Cyprus in 1974, which re-evoked Greek fears about their territorial integrity (Moustakis & Sheehan, 2000: 96). Over the years, hostility remained.[2] After the failed coup d'état attempt in Turkey, in 2016, Greek-Turkish relations further deteriorated, when 8 Turkish officers fleeing from Turkey landed with a military helicopter in Alexandroupolis and asked for asylum. The decision not to deport them back to Turkey led to new tensions between the two countries, with a rapid escalation in March 2018, when two Greek military officers, whilst patrolling Evros border, entered the Turkish territory and were arrested by the Turkish forces. Two years later, in March 2020, a new Greek-Turkish border crisis erupted in Evros (analysed in Chapter 7). Tension continues to mount, with no clear signs of a de-escalation. Rather, in both countries the rhetoric adopted by the media as well as the military and political leadership fuels hostility, preparing the public for an incoming 'hot' episode.

92 *Constructing the border of Evros*

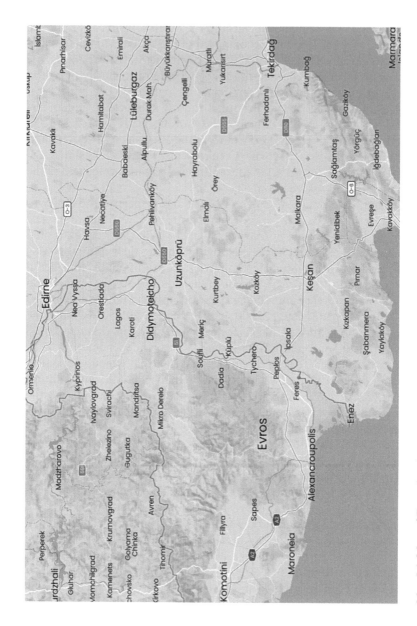

Map 5.1 Map of Evros border.
Source: ©OpenStreetMap, Mapcarta and Mapbox.

The continuous episodes of tension and contention reveal a problematic relationship between the two neighbouring countries, filled with mistrust, suspicion, and rivalry (Moustakis & Sheehan, 2000: 95). Apart from the national context, though, Evros is an EU external border, which both separates and unites the EU and Turkey. It should be noted that Turkey has constituted a country candidate for EU accession since 1999. In spite of the candidate status, many Europeans still consider Turkey an outsider (Düzgit, 2009; Musolff, 2010) rendering Turkey's membership a far distant scenario.

The making of a border

Connecting Greece with Turkey in the northeast and Bulgaria in the northwest, Evros region is marked by substantial cross-border mobility and traffic volume in terms of vehicles, goods, and travellers. In fact, in 2019 more than 750,000 Turks visited Greece (Ethnos, 2020), while more than 900,000 Greeks entered Turkey via Evros land border (Makedonia, 2021).

The Greek-Turkish land border has three border crossing points (BCPs), which manage the entry/exit to Turkey: Kastanies, Kipi, and Pythio. Kipi highway border station is part of Alexandroupolis municipality and it is in the northeast of Evros. It constitutes the busiest crossing point between Greece and Turkey. Crossing this BCP involves, first, document control conducted by border control officers, namely police officers or border guards, and then passing through a bridge over Evros river upon which military officers are stationed. This bridge connects Kipi to the Turkish border station in Ipsala. Meanwhile, Kastanies is part of Orestiada municipality. It is located in the northern part of Evros. BCP Kastanies does not coincide with Evros river. Instead, it is a terrestrial zone containing trees and bushes. As with Kipi, after the border station, there are armed military officers before border crossers reach the neutral zone and then the Turkish border station in Pazarkule. BCP Pythio is the only railway frontier station and rail link between Greece and Turkey. Actually, Pythio was part of the Orient Express' iconic railway route. Geographically it constitutes the easternmost mainland point of Greece and is part of Didymoteicho municipality. The railway lines run in parallel and close proximity to Evros river and a railway bridge connects Greece to Turkey. Similar to Kipi and Kastanies, apart from the BCP, there is also a military station with armed military officers. Mobility through BCPs with document and/or vehicle control constitutes a regular border crossing process. But this is not the sole transborder flow.

Evros has always been identified as a main entry gate for irregular mobility (Frontex, 2007: 15). Yet, like Lampedusa in the 1990s and 2000s, the Greek-Turkish border was not deemed an irregular migration alert. The country's border control spotlight was oriented towards the increased irregular migratory movements on the Greek-Albanian border. In 2009, 40,250 persons, mostly of Albanian origin, crossed the Greek-Albanian border irregularly (Frontex, 2012: 14), whereas only 8,787 used the Evros border

(Hellenic Police, 2022a). Soon this trend changed. Starting in 2010, the number of irregular border crossings in Evros surpassed not only the irregular entries at the Greek-Albanian border but even the 2006 maximum migrant influx of more than 30,000 irregular border crossers in the Canary Islands. In 2010, 47,079 detections were reported at the Greek-Turkish land border, marking a 436% rise. This rise reached its peak in November 2010 with around 350 irregular entries per day (Frontex, 2010a: 8). Proportionately to irregular migration, there was also an ascension in drug trafficking and stolen vehicles (Frontex, 2011a: 33–35). The sudden increase in irregular border crossings was attributed to Turkey's visa liberalisation policy combined with the expansion of Turkish Airlines (Frontex, 2011a: 14–15). That enabled many third country nationals to enter Turkey and then transit irregularly via Evros to Greece before reaching another EU country (Frontex, 2012: 4–5). Indicative is the rise in the number of third country nationals entering Turkey, which from around 5.2 million in 1991 reached 31.4 million in 2011, amounting to a 760% increase in nationals from Egypt, Jordan, Lebanon, and Syria as well as 9,490% rise in Iraqis (Kirişci, 2013: 205–207).

The elevated irregular mobility along the Greek-Turkish land border was accompanied by a proportionate decrease in sea arrivals, manifesting, as a result, a shift in irregular border crossers and facilitators' modus operandi on the Eastern Mediterranean route.[3] Comparing the Greek-Turkish sea and land borders, in the past, most irregular border crossings concerned the Greek-Turkish sea border (see Chart 5.1). However, in 2010 this proportion changed. The strengthening of Frontex's operational activities at the sea border rendered the Greek-Turkish land border more attractive as, after its

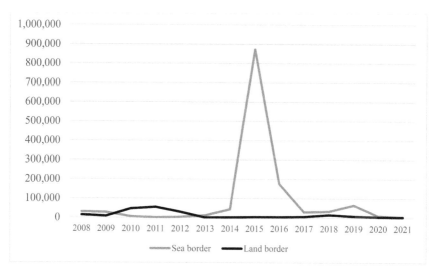

Chart 5.1 Irregular mobility via the Greek-Turkish border.

recent clearance by landmines, smuggling through the land border became cheaper and less risky compared to the sea route (Pallister-Wilkins, 2015: 57).

To cut down the irregular migration flow via the Evros border, in 2012 the Greek authorities erected a 12.5-km long and double barbed-wire-topped fence on the Greek side of the Greek-Turkish land border between the towns of Kastanies in the north and Nea Vyssa in the south. This location was considered a 'vulnerable' point, as it is not delineated by the river (Frontex, 2010a: 8), enabling therefore easy access to the Greek soil.

In spite of the fence, irregular border flow remained high, especially during the migration crisis. In 2015, 3,713 irregular border crossings were reported compared to 1,903 the year before (see Table 5.1). These numbers, though elevated, do not approach the dramatic flow via the sea that reached 872,519 irregular border crossings (Hellenic Police, 2022a). According to Frontex's situational analysis (2014: 133), the reinforcement of the Greek-Turkish land border with surveillance measures and the deployment of additional border guards functioned as a deterrent. Even so, the adoption of the EU-Turkey Statement in March 2016 shifted cross-border mobility once again, as irregular border crossers, to avoid their return to Turkey, opted, rather than the sea route, to cross instead Evros border, which is exempted from the provisions of the Statement.[4] As a result, in 2017, the number of irregular border crossings at the Greek-Turkish land border reached 5,651 detections and tripled the next year.

In 2020, amid the COVID-19 pandemic outbreak, Evros became once again the epicentre of a new border crisis, which induced the installation of new border control tools, such as surveillance cameras and the enhancement of Evros border fence. Hence, irregular mobility at the Greek-Turkish land border has not ceased, despite the Greek-Turkish border being the second most perilous and deadliest migratory route after the Central Mediterranean (Frontex, 2016: 46; Last et al., 2017), with many drownings in the river's waters or deaths due to hypothermia. In parallel, incidents of push-backs to Turkey do not seem to be omitted from Greece's border control tactics, leading to investigations for fundament rights violations (Greek Ombudsman, 2020).

Irregular border crossers, usually with facilitators' help, tend to cross the river at night in groups of approximately 15 to 20 people. They use inflatable boats to disembark onto the Greek bank of Evros river and then on foot they pass through the thick vegetation before moving to Evros' mainland (Greek Council for Refugees, ARSIS & HumanRights360, 2018). The tariff for smugglers' services ranges from €900 up to €2,500 per person (Europol, 2021). Most border crossers are males, mainly from Syria or Afghanistan, although families with children are not a rare occurrence. The 2016 failed coup d'état in Turkey and the purge that followed led to a sudden rise in irregular entries from Turkish nationals, which is quite notable due to the Greek-Turkish rivalry.

Most irregular border crossers that cross the Greek-Turkish land border do not intend to stay in the region of Evros or even in Greece. They aim at reaching another European country, preferably a country of northern

Table 5.1 Irregular migration to Evros

Year	Irregular Border Crossings	Top Nationalities	Causes
2008	14,461	Afghans, Iraqis	• Iraq and Afghanistan wars • Turkey's visa liberalisation • Demining • Turkish airlines expansion
2009	08,787		
2010	47,079	Afghans, Iraqis, Algerians, Moroccans	
2011	54,974	Afghans, Syrians, Algerians	• Arab Spring
2012	30,433		
2013	01,109	Syrians, Afghans	• Arab Winter • Syrian war • Border fence • Sea flows (2014: 43,518≠2012: 3,651)
2014	01,903		
2015	03,713	Syrians, Afghans, Iraqis	• Migration crisis • Coup d'état attempt in Turkey
2016	03,784		
2017	05,651	Syrians, Iraqis, Pakistanis	• EU-Turkey Statement (shift from sea to land border)
2018	15,266	Syrians, Iraqis	• New Greek government
2019	08,642	Afghans, Syrians, Turks, Pakistanis	• COVID-19 pandemic • Border crisis in Evros
2020	05,547		
2021	04,842	Afghans, Syrians	• US withdrawal from Afghanistan & Taliban rule • Border fence extension • Flows towards Cyprus • Migratory pressure towards EU MS neighbouring Belarus

Notes:
Author's elaboration of data.
Data for flows extracted from Hellenic Police (2022a). Data for 2020 and 2021 were provided by the Hellenic Police (Border Protection Division) on 7 April 2022 in the form of a written reply after an official request made by the author.

Europe, either through the Western Balkans land route or by sea through ferries to Italy. Conversely, they try to embark on an intra EU/Schengen flight from a Greek airport with fraudulent documents (Frontex, 2013: 20).

The above indicate that Evros is not just a geographic location or a point drawn on a map. It is part of the EU border regime, attracting each year thousands of border crossers that aspire to reach Europe, while simultaneously it is being shaped by the geopolitical constellations of the Greek-Turkish clash. In this territorial environment, different border control actors partake in Evros' border control conduct, forming a border control 'topos' for the implementation of border control policies and the testing of new border control tools.

Border control conduct and actors

In Greece, border control is in the Police's remit. The Hellenic Police (Astynomia — 'Αστυνομία') is responsible for the management of border control and the protection of national borders (Hellenic Police, 2019). For Greece's land and air borders, this constitutes an exclusive competence of the Hellenic Police. For sea borders, the border control jurisdiction is exerted by both the Hellenic Police and the Hellenic Coastguard. The Army's involvement is subsidiary: assisting the Hellenic Police and the Hellenic Coastguard with border control and surveillance (Interviewee 4). The Hellenic Police is a law enforcement agency under the Ministry of Citizen Protection. The Hellenic Coastguard is oversighted by the Ministry of Maritime Affairs and Insular Police, whereas the Hellenic Army vests in the Ministry of National Defence.

In Evros, the Hellenic Police has two central Police Directorates: in Alexandroupolis for southern Evros and in Orestiada for the northern part. Along the land border, there are border stations[5] for the border control conduct equipped with border guards and technical assets as well as regional and local centres for integrated border and migration management,[6] aiming at maintaining a real-time situational picture for irregular mobility and cross-border crime in Evros. Moreover, in 2012, in Nea Vyssa, an operational border surveillance centre was established that functions as a hub collecting all the optical data from thermal imaging cameras in Evros. The information collected by these centres is turned into statistical data and then analysed before being forwarded to the main centre in Athens, which was inaugurated in 2014 (Hellenic Police, 2014).

For the border control conduct, the Hellenic Police has at its disposal technical means, like vehicles, thermal imaging cameras, day and night cameras, hand-held cameras, binoculars, detector dogs, and heartbeat sensors (Interviewee 4, 5, 6, and 9). Recently, the Greek arsenal became enriched with long range acoustic devices, namely sound cannons, used for deterrence. Moreover, better lighting was installed along the border. Greece has also started using an automatic border surveillance system with sensors, cameras, and radars. The surveillance range is up to 15 km from the border and into the Turkish soil.

The border fence, which constitutes a technical hurdle, is considered another means assisting in border control (Interviewee 4). In 2020, the fence was repaired and reinforced with steel railings, due to severe damage, as, until then, no maintenance work had been undertaken since its 2012 erection (Interviewee 6). Evros fence was extended with new steel constructions that were erected at specific locations in the region of Feres in the northeast of Evros. The fence extension has a total length of 26 km, which has been added to the initial 12.5 km, whereas there are plans for additional an 80 km of fence construction.

Regarding border control personnel, the 'border guard' institution commenced in 1999, having as a task to tackle irregular migration and cross-border crime (Hellenic Police, 2022b). Border guards are part of the Hellenic Police agency structure and therefore not a self-governing and independent corps. Over the years, this has caused qualitative and quantitative frailness of the border control personnel. From the 4,500 border guards recruited in the period 1999–2002, 20 years later almost half, between 2,000 and 2,500, remained serving in border stations (Interviewee 5). Similarly, in Evros, from the initial 400 statutory border guards in 1999, this number fell to 300 (Interviewee 6). Some were sent to serve in other Police Services, while others, due to police personnel shortages, began engaging with tasks irrelevant to border control (Interviewee 5). At the same time, the serving officers started to surpass the age of 40 years (Interviewee 5 and 7). As of 2020, new border guards were recruited, from which 400 were deployed to Evros region, whereas there are plans for additional officers.[7]

To fill staff shortages, the option of police officers' secondment to Evros is often favoured. This practice commenced with Operation 'Shield' in August 2012,[8] when 1,881 border guards and police officers were transferred to Evros (Ministry of Citizen Protection, 2012). Yet, some of the seconded police officers have no previous training or experience in border control. For instance, even officers from the units for the Reinstatement of Order ('MAT') have been transferred to Evros, despite the personnel of these units not being accustomed to dealing with the vulnerable populations often encountered at the borders, like unaccompanied minors. Yet, they agree to their secondment or even propose it, due to financial incentives (Interviewee 5).

While on duty, border guards wear police officer uniforms. Their work schedule involves 8-hour shifts from 06.00 to 14.00, 14.00 to 22.00, and 22.00 to 06.00, for 24-hour cover (Interviewee 5). During these shifts, they conduct foot, vehicle, or boat patrols. Alternatively, they go to uphill areas close to or across the river, for long-distance surveillance, using hand-held cameras or binoculars. Besides, they set up roadblocks in strategic points at the Orestiada-Alexandroupolis highway or on regional roads near the river to dismantle smugglers and irregular migrants. Hence, border control in Evros involves diverse technical means and variant personnel. Greek officers, though, are not the sole border control actors present on the field. There is also Frontex.

Frontex on the field

Ever since 2006, when the agency launched its joint operation 'Poseidon', Frontex has had a continuous presence at the Greek-Turkish land border in Evros coordinating and implementing joint operations almost throughout the year, such as the past 'Poseidon Land' (inaugurated in 2006) or 'Focal Points' (inaugurated in 2008), 'Flexible Operational Activities' (inaugurated in 2015),[9] and 'Terra', the agency's most recent operation in the region (inaugurated in 2022). Frontex's role on the ground involves border patrolling and assistance with detained irregular border crossers in the form of screening and debriefing for intelligence purposes. To do so, it deploys technical equipment and Frontex guest officers or Frontex border guards derived from its standing corps. For instance, currently, in the framework of its joint operation Terra, Frontex is deploying patrol cars and vehicles for border surveillance in Evros as well as around 60 officers from its newly-formed standing corps.[10] During Frontex operations, border patrols are usually performed by a team of two Frontex officers and one Greek border guard (Interviewee 5 and 7).

Aside from joint operations, Evros has twice become the operational theatre of Frontex's rapid interventions. Notably, the first ever RABIT deployment was in Evros. On 24 October 2010, Greece requested the activation of the RABIT mechanism, due to unprecedented irregular migration pressures in Evros (Frontex, 2010b). The RABIT deployment was launched on 2 November 2010 and concluded on 2 March 2011. During the 4-month period, 567 officers from 26 EU member states and Schengen-associated countries arrived in Evros (Frontex, 2011b: 15). Those officers were border guards specialised in false documents, irregular entry, border checks, stolen vehicles, as well as dog handlers, debriefers, and interpreters (Frontex, 2010c). The technical means deployed for the purposes of this operation included fixed-wing aircrafts, helicopters, radars, night vision cameras, buses, patrol cars, and thermo-vision vehicles (Frontex, 2010a: 28). As a result, during the RABIT operation, modern technological assets were deployed at the Greek-Turkish land border as well as considerable professional knowledge and know-how was transferred to Greece (Trauner, 2016: 317). The next RABIT operation in Evros was conducted 10 years later, in the wake of the Evros border crisis.[11] During the 2020 Evros RABIT operation, 327 officers were deployed with 133 patrol vehicles (Frontex, 2021a: 39).

Other Frontex operational activities in Evros include a rapid intervention exercise carried out in 2015, staff exchange operations as well as testing of new border control tools, like aerostat systems in Alexandroupolis for border surveillance (Frontex, 2021b). Frontex's enhanced operational activity in Evros was accompanied by a significant budget growth, which, whilst proportionate to the general rise in the agency's financial resources, manifests Frontex's enhanced land border operational scope. From €2,110,719 in 2007 (Frontex, 2008: 22–23), 10 years later the budget for land operations activities reached the amount of €13,845,000 (Frontex, 2017: 68).

Besides land border activities, in October 2010, just a few days ahead of the launch of its first RABIT deployment in Evros, Frontex created a Frontex Operational Office in Greece, which was renamed as Frontex Liaison Office (FLO) and now EURTF office. Its premises are located inside the Ministry of Maritime and Island Policy in Piraeus and are staffed with 5 persons. This office aims at providing regional support to Frontex activities in a permanent framework, whilst fostering situational awareness.

Apart from the coordination of joint operations at the sea, air, and land borders, Frontex also assists Greece in the implementation of the hotspot approach on the islands by registering, fingerprinting, and screening incoming migrants. Also, it deploys return specialists and organises readmission operations from Greece to Turkey in the framework of the EU-Turkey Statement. In this context, in 2019, 197 third country nationals were readmitted to Turkey (Frontex, 2020: 38).

Beyond Frontex's activity on the Greek soil, the agency has promoted its role on the other side of the border. To this end, it has developed its cooperation with Turkey, a third country neighbouring with two EU members, namely Greece and Bulgaria. Accordingly, in 2009, after Frontex's request, Turkey agreed to appoint a contact point for Frontex (Frontex, 2009: 27). In 2012, Frontex and Turkey signed a Working Arrangement, which enabled Turkey's participation in various air and sea joint operations. Moreover, in 2016 Frontex deployed in Ankara its first liaison officer to a third country. Frontex has also established a Turkey-Frontex Risk Analysis Network (TU-RAN) for risk analysis and strategic planning. Furthermore, Frontex's cooperation with Turkey involves the provision of training to Turkish officers as well as the agency's participation in technical assistance projects for the enhancement of Turkey's migration management capacity and the transfer of best practices.

The above attest Frontex's active role in Evros and generally in Greece since 2006. The agency coordinates and implements land operations, returns irregular migrants to third countries, cooperates with Turkey, brings technical and human resources, such as more than 600 officers (Frontex, 2022b), engages with hotspots, trains Greek border guards (Frontex, 2020: 55) as well as allows Greek officers to gain experience and adapt their work methods by participating in sea, land, and air operations coordinated by Frontex. In fact, in terms of human resources in Frontex joint operations, Greece is listed as the main provider (Frontex, 2020: 87). At the same time, Frontex self-evaluates its operations in Evros as effective, given that they lead to a decrease in the number of irregular entries at this border (Frontex, 2021c).

Yet, some Greek border control officers perceived Frontex negatively, especially during the first years of its operation, as, by recording every irregular migrant that enters Greece, secondary movements to other European countries were prevented, thereby increasing the number of foreign nationals staying in Greece due to the Dublin Regulation[12] (Skleparis, 2016: 99–100). Later though this perspective seems to have been reversed. Greek border officers accept and even ask for Frontex's presence at the Greek-Turkish border (Interviewee 5). Moreover, the Greek Minister of Migration and Asylum (2021) recently honoured Frontex's Executive Director with a medal, manifesting with this symbolic gesture Greece's recognition of Frontex's contribution in dealing with irregular migration flows.

Constructing the border of Evros 101

Thus, besides Greek border control authorities, Frontex is a main border control actor in Evros. Its actorness does not derive solely from its presence at the Greek-Turkish land border. Most importantly, Frontex shapes the border control conduct through its diverse activities and enhanced role in migration management (see Table 5.2). By bringing new assets, deploying

Table 5.2 Frontex in Evros

		Activity	*Scope*	*Composition*
Land Operations		Poseidon	surveillance, migration management, operational coordination	experts (incl. debriefers)
		Focal Points	technical support, operational coordination, border security, staff exchange	document officers, first and second-line officers, stolen vehicle officers, dog handlers
		Flexible Operational Activities	surveillance, cross-border crime, capacity building	debriefers, first and second-line officers, stolen vehicle officers, trainers
		Terra	surveillance, cross-border crime, terrorism, tailored support	surveillance officers, document officers
		RABIT (I & II)	rapid assistance	debriefers, surveillance officers, dog handlers
Other		RABIT exercise		
		Staff exchange		
		Training		
		Hotspots		
		Operational procedures	i.e. biometrics	
		Returns		
		Readmissions	from Greece to Turkey	
		EURTF	regional support, situational awareness, operational management	5-person staff
		External relations with Turkey	working arrangement, technical projects, training, participation in Frontex joint operations	FLO, TU-RAN

officers, coordinating joint operations, or conducting rapid interventions that impact on the migration flows, Frontex frames the territoriality of Evros as a border and its border control policy.

Border control assumptions and practices in Evros

A particular pattern is identified in the conduct of border control in Evros. This pattern refers to a routinised mode for border control, from which it is possible to discern certain cultural traits consisting of background assumptions and practices for the border control conduct (Zaiotti, 2011).

Concerning border assumptions, the main characteristic of Evros is that it constitutes an EU external border and, in parallel, a demarcation of national sovereignty. Being an EU external border, Evros is perceived as an area of responsibility or, put differently, an area of interest for Frontex. On that account, characteristic is the following discourse:

> [Border guard] 'Frontex's headquarters should have been here. Even in Evros. Or at least one of Frontex's operational offices'
>
> (Interviewee 7).

By virtue of its 'EU external border' status, Evros becomes the 'topos' of Frontex operations and therefore the deployment area of various non-Greek border control officers who come to Evros to assist their Greek colleagues in their border control duties. In parallel, the national aspect is shaped by the Greek-Turkish rivalry. As a result, Evros is becoming militarised as a means to ensure territorial integrity and border inviolability (Nachmani, 2003: 172–173; Gkintidis, 2013: 457–458). Therefore, it is not peculiar that the 2020 Evros border crisis was referred to as the 'battle of Evros', attributing to it military characteristics. Furthermore, border militarisation is attested by the fact that, until a few years ago, Evros was a mined area, as Greece, after the 1974 Cyprus crisis, laid landmines as a defensive barrier against a Turkish invasion (Baldwin-Edwards, 2006: 119). Even today, access to the river is restricted. Only military staff and border guards can approach Evros border, namely the borderline and the area near the borderland (Pallister-Wilkins, 2015: 56). The same access restriction applies to the roads and fields that are close to the river, given that metallic chain barriers block vehicles and pedestrians from approaching Evros river. In this militarisation context, the existence of military stations next to the Greek-Turkish BCPs comes as no surprise. Moreover, more than 10 military camps operate in villages near the river, such as in Feres, Peplos, Kavyli, Tychero, and Didymoteicho.

Turning to border control assumptions, border control in Evros unfolds a securitisation context, as migration is considered a security concern for border control practitioners. In this spirit, border officers are being preoccupied by security threats, such as terrorism, as shown below:

> [Border guard] 'The biggest threat is terrorism, namely the risk of terrorists entering the country'
>
> (Interviewee 5).

Or consider irregular migration a security anxiety for the whole Schengen area:

> [Police officer] 'The problem with irregular migration is that it challenges Schengen's internal security. Schengen's internal security becomes inadequate'
>
> (Interviewee 9).

Securitisation is empirically manifested with the fence construction (Skleparis, 2016: 96; Grigoriadis & Dilek, 2019) or with Frontex's operations conducted in Evros, like the activation of RABIT during an emergency (Léonard, 2010: 244–245).

Moreover, a technocratic approach has been put forward that emphasises specialised knowledge, as manifested in the following discourses:

> [Police officer] 'Frontex had the expertise but we had the experience [...] We have taken the expertise from Frontex'
>
> (Interviewee 9).

> [Border control officer] 'We have acquired the expertise and improved both our staff and processes'
>
> (Interviewee 4).

> [Border guard] 'Trainings for screeners and debriefers are really important [...] We now have our own experts after being trained by Frontex'
>
> (Interviewee 5).

Apart from expertise and experts, technocracy is also fostered with the deployment of high-tech and computer-assisted border control tools, such as IT reporting systems. In fact, on a daily basis, Greek authorities use Frontex applications like JORA and EUROSUR to upload border control incidents (Interviewees 6, 7, and 9).

Another characteristic is that the 'topos' of border control has been shifted, redefining the internal/external distinction. Border control is not solely carried out at the border but away from its actual geographic location, creating new border control spaces (Walters, 2006: 193). It has been expanded beyond the border and into the territory of the third country, accounting for an extra-territorial strategy. The importance accorded to extra-territorialisation is evident in the following extract:

> [Police Major in Eastern Macedonia and Thrace] 'Our operational strategy worked because we had the early pre-frontier picture that something massive was happening on the other side of the border'
> (Kathimerini, 2020).

Border control in the pre-frontier area occurs, for instance, with the use of long-range surveillance devices, such as radars and aerostats, allowing the collection of visual information and therefore border patrolling within the Turkish soil. Yet, simultaneously, border control is being transferred inside the country's territory. Setting up roadblocks in highways or other regional roads to dismantle smugglers and irregular migrants, after crossing the border and transporting within Greece, denotes a parallel intra-territorialisation of border control by moving its conduct within the country and away from the border. In this spirit, a border guard mentions that 'we care what is happening after [irregular migrants' entry], that is after the border [...] [Irregular migrants] leave from the location of the border and can reach even Thessaloniki [...] We set up roadblocks after their entry to stop them' (Interviewee 5).

In Evros, it is possible to discern the increasing use of and dependence on border surveillance. Actually, border surveillance is being considered as a major border control goal, as attested in the following:

> [Border control officer] 'What do you want from border control? The surveillance of the borders so as to have adequate time to react'
> (Interviewee 4).

Surveillance's enhancement is being undertaken with the deployment of surveillance officers or diverse technical means, like fixed-wing aircrafts, helicopters, radars, and patrol cars. Furthermore, an automatic border surveillance system has been installed, which was recently further expanded, consisting of monitoring stations, thermal sensors, cameras, radars, as well as system monitoring and surveillance of surrounding areas (closed-circuit television — CCTV). Surveillance is also fostered by the border surveillance centre in Nea Vyssa, which checks all the cameras installed. Furthermore, surveillance is integrated as an operational aim in Frontex joint operations in Evros.

Moreover, there is a move towards an intelligence-oriented approach to border control. Illustrative of that are the following:

> [Police officer] 'Very important is the information that we collect from the apprehended irregular migrants about smuggling networks'
> (Interviewee 9).

[Border control officer] 'The collection of information is horizontal [...] Situational awareness is of key importance'
(Interviewee 4).

The emphasis on intelligence is being manifested through debriefing activities, which are included in almost every Frontex joint operation conducted at the Greek-Turkish land border. Intelligence is also collected via the TU-RAN risk analysis network established between Frontex and Turkey. An intelligence logic is also manifested through the establishment of integrated border and migration management centres, given that they analyse all the information that they collect in order to maintain situational awareness and proactively identify possible risks and emerging threats.

Regarding border control practices produced at the border, Evros is based on a policing border control organisation focusing, as a result, on the detection and prevention of criminal activities (Mawby, 2011: 17). Police officers and border guards, who are part of the Police, carry out border controls, whilst wearing police uniforms. At the same time, policing tactics are commonly employed for border control purposes. These include boat, vehicle, and foot patrols, debriefing, roadblocks, and the use of dogs to dismantle migrants or prohibited goods. It should be noted that debriefing as well as dog and boat patrolling were both initiated in the border of Evros by Frontex (Interviewee 6 and 7).

Another practice is information gathering. The collection, sorting, visual representation, and exchange of information have become dominant border control activities in Evros, implemented mainly via the integrated border and migration management centres and the operational border surveillance centre. All the data collected are also sent to Frontex, consolidating and fostering this information gathering and exchanging environment, which is also promoted by the EURTF office in Piraeus.

Moreover, border control in Evros is being conducted and relies on a practice of multilateral cooperation. Greece, Frontex, and even Turkey all cooperate to manage irregular migration flows at the Greek-Turkish land border. In this spirit, Greek border authorities have frequent meetings and web conferences with officers from Frontex's headquarters to discuss migratory challenges (Interviewee 3). During the inaugural RABIT operation, Frontex, Greek, and Turkish border officials held frequent informal meetings on both sides of the land border, trying to advance operational communication and synergy among relevant actors (Frontex, 2010a: 10–11). Furthermore, on the ground, officers from different authorities and member states are being deployed in Evros, demonstrating a prism of multilateral cooperation. Working multilaterally, and in spite of any communication difficulties due to language barriers (Interviewee 5), these officers build and maintain interpersonal relations during and after their deployment at Evros border (Interviewee 6, 7, 8, and 9). They share their experiences, discuss work challenges, confront similar problems, and even set common

106 *Constructing the border of Evros*

professional goals, establishing a connection with their colleagues. This describes the development of a border control community that exists and is being nourished at the border.

Technologically enhanced functions and procedures constitute another practice encountered extensively across Evros. High-tech border control tools, such as automated surveillance systems, sophisticated visualisation devices, thermal imaging cameras, biometrics, thermo-vision vans, and heartbeat detectors are all encountered in Evros and are being used for border control purposes. Likewise, Evros functions as a testing area for innovative border control solutions, as shown with Frontex's trials of aerostat systems in Alexandroupolis.

Professionalisation is another border control practice traced in Evros. Actually, border guards voice that they 'love [their] profession' (Interviewee 5). Professionalisation stems from the spread of seconded Greek officers and Frontex experts deployed in Evros. Experts, namely professionals, such as debriefers, trainers, dog handlers, surveillance officers, all arrive in Evros to provide their professional expertise and know-how. At the same time, the economic incentive for officers' transfer to this border enhances professionalism, rendering their assignment to Evros a profitable job. Professionalisation is also advanced through the provision of specialised training. In fact, Frontex's Common Core Curriculum for border guard training has been incorporated into Greece's police training, whereas Frontex has developed specialised curricula for land borders, such as training for land border surveillance officers.

The above elements of border control represent Evros' border control assumptions and practices. These assumptions and practices frame the border control conduct in this border, implemented and reproduced by the border control practitioners. Being integrated in the border control implementation and therefore displaying a routinised border control mode, these assumptions and practices constitute cultural traits for the border control conduct at Evros border (see Table 5.3).

Table 5.3 Border control cultural traits in Evros

Border assumptions	EU external border – post-national integration
	National territory – militarisation
Border control assumptions	Securitisation
	Extra-territorialisation / Intra-territorialisation
	Technocracy
	Surveillance
	Intelligence
Border control practices	Policing
	Information gathering
	Multilateral cooperation
	Technology
	Professionalisation

Evros and Lampedusa: drawing differences and similarities

The comparative analysis of both borders allows the deduction of conclusions about their function and development as well as the border control mode pursued for their effective 'protection'. Turning to the comparative research results, it is possible to discern that Evros and Lampedusa, although they are different borders, share certain common elements.

In general, Evros and Lampedusa constitute EU external borders that record high migratory flows. To manage these flows, Frontex has built almost a continuous presence on both borders, functioning as a key border control actor and facilitating the development of a community of border control practitioners. Actually, the range and type of Frontex's activities on both borders are noteworthy, including, for example, joint operations, assistance in hotspots, risk analysis, testing of new technologies, as well as the deployment of certain common expert profiles, such as debriefing officers.

Yet, despite intense migratory pressures and becoming hosting places for Frontex's joint operations, Evros and Lampedusa differ substantially as borders. The main difference emanates from their geographic characteristics, given that Evros is a land border, whereas Lampedusa is a sea border and an island. In the case of a land border, the border-line becomes a tangible element enabling even the erection of a barbed-wire fence. For a sea border, though, there are no concrete lines demarcating where one country begins and the other ends. Sea/land variation results in the use of different tools for border control, such as boats for Lampedusa but dogs for Evros patrols, as well as variant human resources, like SAR experts in Lampedusa and stolen vehicle experts in Evros. Likewise, the type of border control differs; in Lampedusa emphasis is placed on SAR, in Evros on the deterrence of irregular entries or, when not possible, on their apprehension with the use of roadblocks away from the border. Moreover, Frontex's RABIT mechanism has been activated twice in Evros, but never in Lampedusa, although a state of emergency has been declared on the island at various times. Another variance relates to the third countries on the other side of the border, that is, Turkey in the case of Evros and North African countries, namely Libya and Tunisia, for Lampedusa. Although bilateral tensions had an impact on both borders, in the case of Evros, the context of never-decreasing Greek-Turkish enmity severely affects its border management, leading to a militarised border environment.

Comparing their cultural traits, these two borders display similar and variant assumptions and practices. Accordingly, on border assumptions, the 'EU external borders' title leads both borders to be perceived as part of the EU. This perception is being reflected and advanced with the deployment of border guards from other European countries there during Frontex operations, which manifests a spirit of post-national integration and burden-sharing among the EU countries. In parallel, both borders maintain a strong national context, which defines them as symbols of the state's territory. On that account, borders need to be safeguarded and effectively controlled to avoid any unauthorised or, put differently, unwanted entries to the country. The national aspect drives both borders to be militarised, yet diversely. Evros militarisation is a product of the Greek-Turkish hostility, as Greeks consider Turkey a country threatening their territorial integrity and therefore maintain in Evros significant military forces to act in case of an armed attack. In Lampedusa, though, the border has become militarised due to the border control approach implemented there, as it involves mostly military means and forces.

On border control assumptions, both borders are securitised, dominated by conceptions of intelligence, surveillance, technocracy, and sharing a common understanding regarding the re-territorialisation of borders and border control by shifting them away from borders' geographic location. A key variance relates to militarisation. Although Evros border is militarised, militarisation has not yet been diffused to its border control conduct. Police officers and border guards, which are part of the Police, are the main national actors implementing border control. The Army's role is still auxiliary, whenever needed, having as primary mission the protection of Greece's territorial integrity against an invasion from Turkey. On the contrary, in Lampedusa almost all national border control actors are part of the military corps, leading therefore to an inevitable militarisation of the border control conduct.

Considering border control practices, on both borders border control has been professionalised and is being implemented via information gathering, multilateral cooperation, and the use of technology. In Evros, though, these border control practices are being accompanied by policing, as national border control officers belong to the Police and therefore are accustomed to policing tactics, which they utilise to intercept irregular border crossers. By contrast, in Lampedusa, border control involves SAR operations. Being a sea border, SAR incidents have become an operational routine for the border control officers there, as most irregular migrants landing on Lampedusa have arrived at the island after being rescued in the context of a SAR operation.

In broad summation, Evros and Lampedusa are clearly two different EU external borders in geographic, geopolitical, symbolic, and functional terms. Regardless of their inherent differences, they still share certain mutual elements, and particularly certain common assumptions and practices for border control. Yet, the presence of those common cultural traits was unexpected due to the differences of the borders. So, identifying these common assumptions and practices on two different borders indicates that they are not random border pieces or national-defined characteristics. Rather, being cultural traits for the border control conduct, they may be components of an emerging border control culture.

Conclusion

Concluding the comparative case study analysis, which started in Chapter 4 with Lampedusa border, this chapter explored the Greek-Turkish land border of Evros. Evros border attracts a large number of irregular border crossers that want to reach Europe using it as a stepping stone towards their final settlement in a northern European country. As a result, it has become the field of various Frontex operational activities, since the agency's operational launch. Aside from being a migration hotspot, the Greek-Turkish border of Evros is a territorial manifestation of the problematic or even hostile relationship between these two neighbouring states.

Drawing on this context, the chapter sought to explore Evros' border control characteristics and cultural traits. Maintaining the same structure as Lampedusa's chapter so as to facilitate the comparison and the extraction of conclusions, this chapter firstly analysed Evros as a border, including its geographical characteristics, its border development, and border control function. Subsequently, it investigated the border control approach implemented there, extracting cultural traits, namely border control assumptions and practices. The chapter argued that Evros has been transformed as a border. Its transformation refers to its dual constitution as both a national and an EU external border. Through Evros' analysis, the chapter also showed that Frontex is a border control actor. Frontex's actorness is not limited to the agency's operational presence at the border. Frontex also actively participates in the border control conduct, promoting some of the cultural traits encountered there.

Regarding the comparative case study conclusion, the chapter's final section summarised the research results, highlighting the differences and similarities of Evros and Lampedusa. Actually, the chapter revealed that there are several common border control assumptions and practices in Evros and Lampedusa despite their differences, both as borders and border control 'theatres'. Put differently, there are common cultural traits in these two different borders that point towards the emergence of a (new) EU border control culture. Indeed, this is the central theme of the next chapter. Drawing on the comparative case study's results as well as the analysis on Frontex, it will investigate whether the assumptions and practices traced are not just cultural traits but instead parts of a border control culture currently pursued.

Notes

1 Apart from Greece, the EU border with Turkey also involves the Bulgarian-Turkish land border.
2 For confrontation issues between Greece and Turkey, see Moustakis and Sheehan (2000).
3 The Eastern Mediterranean sea route refers to irregular migratory flows towards Greece, Cyprus, and Bulgaria.

4 The EU-Turkey Statement foresees that all migrants who arrived on Greek islands via Turkey after 20 March 2016 will be returned to Turkey, as Turkey is declared a safe third country (Council of the EU, 2016).
5 These stations have been established in the towns of Soufli, Tychero, Feres, Orestiada, Didymoteicho, Metaxades, and Kyprinos.
6 Regional centres in the towns of Alexandroupolis and Neo Cheimonio as well as local centres in Soufli, Tychero, Feres, and Didymoteicho.
7 In the coming days, after finalising their recruitment, 250 new border guards are to be deployed at Evros border.
8 For more information on operation Shield (Aspida), see Baird (2017: 75–76).
9 The operation 'Flexible Operational Activities' was inaugurated in 2014 at the borders of Hungary, Croatia, and Poland. The next year, its geographic focus was amended to include the land borders of Croatia, Bulgaria, Hungary, and Greece.
10 Data provided by Frontex on 10 February 2022 in the form of a written reply after an official request made by the author.
11 Besides Evros, the RABIT mechanism has also been activated twice at Greece's sea borders in the Aegean (2015 and 2020) as well as once at Lithuania's land border with Belarus (2021).
12 The Dublin Regulation lays down the criteria to be applied when determining the country responsible for examining an application for international protection. In most cases the responsibility is allocated to the EU country the asylum seeker first entered.

References

Argyrou, V. (2006) 'How Greeks Think: About Turks, for Example'. *South European Society & Politics*, 11(1), pp. 33–46.

Baird, T. (2017) *Human Smuggling in the Eastern Mediterranean*. Abingdon: Routledge.

Baldwin-Edwards, M. (2006) 'Migration between Greece and Turkey: From the Exchange of Populations to the Non-Recognition of Borders'. *South-East Europe Review for Labour and Social Affairs*, 9(3), pp. 115–122.

Çarkoğlu, A. & Kirişci, K. (2004) 'The View from Turkey: Perceptions of Greeks and Greek-Turkish Rapprochement by the Turkish Public'. *Turkish Studies*, 5(1), pp. 117–153.

Christianou, S. (2009) 'Prefecture of Evros: Natural Environment'. Available at http://www.xanthi.ilsp.gr/cultureportalweb/article.php?article id=1065&topic_id=15&level=1&belongs=0&area_id=2&lang=en (accessed March 2022).

Council of the EU (2016) 'EU-Turkey Statement: 18 March 2016'. Available at https://www.consilium.europa.eu/en/press/press-releases/2016/03/18/eu-turkey-statement/ (accessed March 2022).

Düzgit, A. (2009) 'Constructing Europe Through Turkey: French Perceptions on Turkey's Accession to the European Union'. *Politique Européenne*, 3(29), pp. 47–82.

Ethnos (2020) 'Newsroom: Alexandroupolis'. Available in Greek at https://www.ethnos.gr/greece/article/82467/alexandroypolhpanoapo755000toyrkoiepiskefthh kanthnelladato2019 (accessed March 2022).

European Commission (2020) 'Remarks by President von der Leyen at the Joint Press Conference with Kyriakos Mitsotakis, Prime Minister of Greece, Andrej Plenković, Prime Minister of Croatia, President Sassoli and President Michel'. Available at https://ec.europa.eu/commission/presscorner/detail/en/statement_20_380 (accessed February 2022).

Europol (2021) 'Nine Arrests in Hit against Network Smuggling Migrants via the Greek-Turkish Land Border'. Available at https://www.europol.europa.eu/media-press/newsroom/news/nine-arrests-in-hit-against-network-smuggling-migrants-greek-turkish-land-border (accessed March 2022).

Frontex (2007) *Frontex General Report 2007*. Warsaw: Frontex.

Frontex (2008) *Frontex General Report 2008*. Warsaw: Frontex.

Frontex (2009) *Frontex General Report 2009*. Warsaw: Frontex.

Frontex (2010a) *Frontex General Report 2010*. Warsaw: Frontex.

Frontex (2010b) 'Frontex Deploys Rapid Border Intervention Teams to Greece'. Available at https://frontex.europa.eu/media-centre/news/news-release/frontex-deploys-rapid-border-intervention-teams-to-greece-PWDQKZ (accessed March 2022).

Frontex (2010c) 'Frontex to Deploy 175 Specialist Border Personnel to Greece'. Available at https://frontex.europa.eu/media-centre/news/news-release/frontex-to-deploy-175-specialist-border-personnel-to-greece-9neidF (accessed March 2022).

Frontex (2011a) *Annual Risk Analysis 2011*. Warsaw: Risk Analysis Unit.

Frontex (2011b) *General Report 2011*. Warsaw: Frontex.

Frontex (2012) *Annual Risk Analysis 2012*. Warsaw: Risk Analysis Unit.

Frontex (2013) *Annual Risk Analysis 2013*. Warsaw: Risk Analysis Unit.

Frontex (2014) *Frontex Programme of Work 2015*. Warsaw: Frontex.

Frontex (2016) *Risk Analysis for 2016*. Warsaw: Risk Analysis Unit.

Frontex (2017) *Programming Document 2018-2020*. Warsaw: Frontex.

Frontex (2020) *Consolidated Annual Activity Report 2019*. Warsaw: Frontex.

Frontex (2021a) *Annual Implementation Report 2020*. Warsaw: Frontex.

Frontex (2021b) 'Frontex Tests Aerostat Systems in Greece for Border Surveillance'. Available at https://frontex.europa.eu/media-centre/news/news-release/frontex-tests-aerostat-systems-in-greece-for-border-surveillance-b5E9I8 (accessed March 2022).

Frontex (2021c) 'Migratory Situation November'. Available at https://frontex.europa.eu/media-centre/news/news-release/migratory-situation-november-the-highest-number-of-detections-in-november-since-2015-Vn2CSr (accessed March 2022).

Frontex (2022a) 'Information Management'. Available at https://frontex.europa.eu/we-know/situational-awareness-and-monitoring/information-management/ (accessed January 2022).

Frontex (2022b) 'Operation Poseidon'. Available at https://frontex.europa.eu/we-support/main-operations/operation-poseidon-greece-/ (accessed March 2022).

Gkintidis, D. (2013) 'Rephrasing Nationalism: Elite Representations of Greek-Turkish Relations in a Greek Border Region'. *Southeast European and Black Sea Studies*, 13(3), pp. 455–468.

Greek Council for Refugees, ARSIS & HumanRights360 (2018) 'The New Normality: Continuous Push-Backs of Third Country Nationals on the Evros River: Report'. Available at https://www.gcr.gr/en/ekdoseis-media/reports/reports/item/1028-the-new-normality-continuous-push-backs-of-third-country-nationals-on-the-evros-river (accessed March 2022).

Greek Ombudsman (2020) 'Alleged Pushbacks to Turkey'. Interim Report.

Grigoriadis, I.N. & Dilek, E. (2019) 'Securitizing Migration in the European Union: Greece and the Evros Fence'. *Journal of Balkan and Near Eastern Studies*, 21(2), pp. 170–186.

Hellenic Police (2014) 'I.B.M.M.C'. Available in Greek at http://www.astynomia.gr/images/stories/2014/prokirikseis14/04062014-kodisme.pdf (accessed March 2022).

Hellenic Police (2019) 'Hellenic Police'. Available in Greek at http://www.astynomia.gr/index.php?option=ozo_content&perform=view&id=1831&Itemid=528&lang= (accessed March 2022).

Hellenic Police (2022a) 'Statistical Data: Illegal Migration'. Available in Greek at http://www.astynomia.gr/index.php?option=ozo_content&lang=&perform=view&id=93710&Itemid=2443&lang= (accessed March 2022).

Hellenic Police (2022b) 'Border Guards'. Available at http://www.astynomia.gr/index.php?option=ozo_content&perform=view&id=56&Itemid=618&lang=EN (accessed March 2022).

Hellenic Statistical Authority (2021) 'Press Release'. Available in Greek at https://www.google.com/url?sa=t&rct=j&q=&esrc=s&source=web&cd=&cad=rja&uact=8&ved=2ahUKEwijmpCliN_1AhUsif0HHVmwBU4QFnoECAUQAQ&url=https%3A%2F%2Fwww.statistics.gr%2Fdocuments%2F20181%2F0dab1fe7-e6ef-ee87-0abc-030a11b74d27&usg=AOvVaw2bDcIvojRY54HYDoKKlRrZ (accessed March 2022).

Heraclides, A. (2012) 'What Will Become of Us without Barbarians? The Enduring Greek-Turkish Rivalry as an Identity-Based Conflict'. *Southeast European and Black Sea Studies*, 12(1), pp. 115–134.

Kathimerini (2020) 'Evros: In the Forbidden Zone'. Available in Greek at https://www.youtube.com/watch?v=5egIsXpJ6e4 (accessed March 2022).

Kirişci, K. (2013) 'The EU, Turkey, and the Arab Spring: Challenges and Opportunities for Regional Integration'. In: S. Aydın-Düzgit, A. Duncker, D. Huber, E.F. Keyman & N. Tocci (eds.), *Global Turkey in Europe: Political, Economic, and Foreign Policy Dimensions of Turkey's Evolving Relationship with the EU*. Rome: Istituto Affari Internazionali, pp. 195–220.

Last, T., Mirto, G., Ulusoy, O., Urquijo, I., Harte, J., Bami, N., Pérez Pérez, M., Macias Delgado, F., Tapella, A., Michalaki, A., Michalitsi, E., Latsoudi, E., Tselepi, N., Chatziprokopiou, M. & Spijkerboer, T. (2017) 'Deaths at the Borders Database: Evidence of Deceased Migrants' Bodies Found Along the Southern External Borders of the European Union'. *Journal of Ethnic and Migration Studies*, 43(5), pp. 693–712.

Léonard, S. (2010) 'EU Border Security and Migration into the European Union: Frontex and Securitisation through Practices'. *European Security*, 19(2), pp. 231–254.

Makedonia (2021) 'Greeks and Bulgarians in the Markets of Turkey'. Available at Greek at https://www.makthes.gr/stis-agores-tis-toyrkias-ellines-kai-voylgaroi-gia-psonia-akoma-kai-gia-akinita-492116 (accessed March 2022).

Mawby, R.I. (2011) 'Models of Policing'. In: T. Newburn (ed.), *Handbook of Policing*. Abingdon: Routledge, pp. 17–46.

Mill, J.S. (1865 [1843]) *A System of Logic: Ratiocinative and Inductive*. London: Longmans.

Millas, H. (2004) 'National Perception of the "Other" and the Persistence of some Images'. In: M. Aydın & K. Ifantis (eds.), *Turkish-Greek Relations: The Security Dilemma in the Aegean*. London: Routledge, pp. 53–66.

Ministry of Citizen Protection (2012) 'Parliamentary Reply'. Available in Greek at www.hellenicparliament.gr/UserFiles/67715b2c-ec81-4f0c-ad6a-476a34d732bd/7696899.pdf (accessed March 2022).

Ministry of Migration & Asylum (2021) 'Award of Commendation Medal to the Executive Director of FRONTEX'. Available at https://migration.gov.gr/en/aponomi-diamnimoneysis-ston-ektelestiko-dieythynti-tis-frontex-fabrice-leggeri/ (accessed March 2022).

Moustakis, F. & Sheehan, M. (2000) 'Greek Security Policy after the Cold War'. *Contemporary Security Policy*, 21(3), pp. 95–115.

Musolff, A. (2010) 'The Eternal Outsider? Scenarios of Turkey's Ambitions to Join the European Union in the German Press'. In: L. Saric, A. Musolff, S. Manz & I. Hudabiunigg (eds.), *Contesting Europe's Eastern Rim: Cultural Identities in Public Discourse*. Bristol: Multilingual Matters, pp. 157–172.

Nachmani, A. (2003) *Turkey Facing a New Millennium: Coping with Intertwined Conflicts*. Manchester: Manchester University Press.

Newman, D. (2003) 'On Borders and Powers: A Theoretical Framework'. *Journal of Borderlands Studies*, 18(1), pp. 13–25.

Newman, D. (2011) 'Contemporary Research Agendas in Border Studies: An Overview'. In: D. Wastl-Water (ed.), *The Ashgate Research Companion to Border Studies*. Farnham: Ashgate, pp. 33–47.

Pallister-Wilkins, P. (2015) 'The Humanitarian Politics of European Border Policing: Frontex and Border Police in Evros'. *International Political Sociology*, 9(1), pp. 53–69.

Skleparis, D. (2016) '(In)securitization and Illiberal Practices on the Fringe of the EU'. *European Security*, 25(1), pp. 92–111.

Theodossopoulos, D. (2006) 'Introduction: The "Turks" in the Imagination of the "Greeks". *South European Society & Politics*, 11(1), pp. 1–32.

Topak, Ö.E. (2014) 'The Biopolitical Border in Practice: Surveillance and Death at the Greece-Turkey Boderzones'. *Environment and Planning D: Society and Space*, 32(5), pp. 815–833.

Trauner, F. (2016) 'Asylum Policy: The EU's "Crises" and the Looming Policy Regime Failure'. *Journal of European Integration*, 38(3), pp. 311–325.

van Houtum, H., Kramsch, O. & Ziefhofer, W. (eds.) (2005) *B/ordering Space*. Aldershot: Ashgate.

Walters, W. (2006) 'Border/Control'. *European Journal of Social Theory*, 9(2), pp. 187–203.

Zaiotti, R. (2011) *Cultures of Border Control: Schengen & the Evolution of European Frontiers*. Chicago: University of Chicago Press.

Interviews

Interviewee 4: Border control officer at Hellenic Police's headquarters, 11 April 2018.
Interviewee 5: Border guard that had been seconded in Evros, 16 October 2018.
Interviewee 6: Border guard in northern Evros, 23 October 2018.
Interviewee 7: Border guard in southern Evros, 23 October 2018.
Interviewee 8: Border guard in Evros, 23 October 2018.
Interviewee 9: Police officer in Evros, 24 October 2018.

6 Border control in process
The rise of Warsaw culture

> 'I see Frontex promoting a working culture and the feeling that all border guards belong to the same community [...]'.
> —Fabrice Leggeri, former Frontex Executive Director
> (Frontex, 2015a)

Introduction

As the above words show, the first day of taking the helm of Frontex, its newly appointed Executive Director talked about a community of border guards and a working culture promoted by Frontex. But what if Frontex is not promoting just a working culture but rather a culture of border control? And what if this culture is not the Schengen culture but a totally new trajectory? What might this mean for the agency and the border control conduct in Europe?

The research at the borders (Chapters 4 and 5), along with Frontex's analysis (Chapters 2 and 3), uncovered certain common assumptions and practices for the border control conduct. Their detection not only in different border locations but also within a central border control actor, that is, Frontex, signifies that they are not idiosyncratic characteristics emerging in a particular geographic area or institutional context. Instead, they are components of a distinctive border control culture that defines the EU border control conduct.

Synthesising and drawing on the previous chapters' research results, this chapter establishes the emergence of the 'Warsaw' border control culture. Comparing it with the other cultures of border control, namely 'Schengen', 'Westphalia', and 'Brussels' (Zaiotti, 2011), it points out that the culture traced constitutes a new and unique border control trajectory, which has already been successfully institutionalised as a border control regime. The rest of the chapter traces and assesses Frontex's impact. It shows that Frontex has promoted and produced the components of this culture, rendering it the new dominant culture for border control in Europe. This finding is mostly based on data from document analysis, institutional discourse

DOI: 10.4324/9781003230250-6

analysis, interviews with Frontex staff and national border officers, as well as evidence from process-tracing and the empirical research at the borders.

The chapter starts with an exploration of the new culture, shedding light on its constitutive elements, namely its assumptions and practices as well as the border control community that pursues it. The next section compares this new culture, labelled by this book as 'Warsaw', with the other border control cultures in Europe (Schengen, Westphalia, and Brussels). Furthermore, it examines the sequential path for cultural evolution from Schengen to Warsaw. Subsequently, it assesses Frontex's role in Warsaw's emergence by formulating predictions about Frontex's impact on Warsaw's constitutive elements, or in other words, Warsaw's essential conditions, namely Warsaw's border control community, Warsaw's reference texts as well as Warsaw's assumptions and practices. The last part discusses the bigger picture, that is, the role of the EU institutions in Warsaw's consolidation. The chapter concludes that Frontex is responsible for the emergence and (re)production of the components that constitute this new border control culture. Thus, Frontex is not just an agent or instrument of EU border control. It is way more than that and, therefore, way more than what it was originally designed and anticipated by its creators to be.

Culture loading

According to Zaiotti (2011: 23), a culture of border control consists of assumptions and practices shared by a border control policy community in a given period. Based on this, the research at the borders (Chapters 4 and 5) and Frontex (Chapter 3) sought to extract border control assumptions and practices in order to investigate the border control culture currently pursued. It revealed that there are common assumptions and practices not only in different borders, namely Lampedusa and Evros, but also in a key border control actor, that is, Frontex. The common elements traced compose and signify a new border control culture that exists at Europe's borders. The next step is to zoom in on this culture to analyse its components, namely its assumptions and practices.

Regarding border assumptions, borders are considered both national and EU external borders. This denotes borders' binary nature and function. On the one hand, borders delimitate national sovereignty, as member states, being sovereign states, 'continue to keep their competence and sovereignty over their borders' (European Commission, 2016a). On the other, they represent a common EU external border built upon the ideal of supranational integration. EU borders' binary representation leads to the management of external borders '[i]n a spirit of shared responsibility' between Frontex and national authorities (Regulation, 2019), consolidating therefore Frontex's role as a central border control actor.

116 *Border control in process*

Turning to the underlying border control assumptions, these include securitisation, technocracy, extra-territorialisation/intra-territorialisation, surveillance, and intelligence. All of them were traced both at the borders and Frontex.

Starting with securitisation, border control is securitised, because irregular migration is framed as a security problem or threat that must be combatted (Council of the EU, 2018). In this context, irregular migrants are perceived as a source of insecurity threatening both the national state and the deepening of the EU project. To protect the latter, additional funds as well as more border control officers and new border control tools are being deployed at the border, having as an aim more effective border controls and leading, in turn, to the tightening of the EU borders and the conception of borders as frontiers separating non-EU citizens from EU citizens. So far, such framing has steered to the adoption of emergency measures both at national and EU level, like the reintroduction of internal border control checks within the Schengen area 'in exceptional circumstances' (Regulation, 2016b) or the development of new policy tools as a migration crisis response, like the hotspot approach (European Commission, 2015a). In parallel, by linking migration to security, border control officers and Frontex become guardians of national and EU security, overriding therefore their technical border guarding functions.

Border control has acquired an advanced technocratic dimension. Border control manifests a managerial and technocratic approach with border control practitioners adopting common operational responses and pursuing similar strategies. Actually, border control is being conducted via specialised technological tools, for instance, EUROSUR and ETIAS computer systems. Emphasis is also placed on technocratic expertise, namely experts that have developed specialised knowledge and technical skills, such as return experts, debriefers, and screeners (Regulation, 2016a; Frontex, 2021). The development and inclusion of technocratic tools in border control does not solely manifest a technocratic spirit in EU border control. Technocracy also evokes standardised responses by the border control practitioners. Using the same platforms and computer systems as well as being trained through the same curricula or educated in common courses and creating similar expert profiles, border control practitioners start producing patterned border control behaviours.

Border control reveals a simultaneous, yet conflicting, extra-territorialisation, and intra-territorialisation. Its conduct does not take place at the borderline. Rather, border control activities are performed beyond and inside the border, shifting in turn the space of the border both within and outside the borderline (Regulation, 2013; Council of the EU, 2018). Thus, the conduct of border control is being dissociated from the location of the border, entitling as a result border control actors to transfer their border control tasks away from the territory of the EU external border, expanding therefore both their duty station and their field tasks. Actually, as borderlines lose their geographic relevance, border control officers or border control actors, accept the idea of transforming into control officers or control actors tasked to control rather than to control the border.

Border control also increasingly emphasises the input of intelligence. For instance, the need to acquire intelligence on migratory routes, intended travellers as well as developments at local, regional or international level that may impact cross-border mobility. That is, the collection and use of intelligence so as to act in a 'proactive way' (European Commission, 2015b) and for the 'proactive management of migration' (Regulation, 2019). The aim is to maintain an up-to-date situational picture as well as an improved understanding of the wider environment that permits the identification of possible risks that may evolve into security threats or border challenges. This intelligence-led environment is fostered with the establishment of risk analysis structures and intelligence-sharing networks, like the Western Balkans Risk Analysis (WB-RAN) or the deployment of liaison officers both in EU member states and third countries.

Lastly, border control is characterised by and rests upon surveillance. Border surveillance refers not just to the monitoring of the border. Instead, it involves a strategic decision, which links vision with action, as after seeing, border control officers act to prevent unauthorised entries (Dijstelbloem et al., 2017). Constant surveillance is ensured with the installation of sophisticated monitoring tools along the EU external border, such as cameras and radars. Not surprisingly, border control is not merely defined as border checks; it also encompasses border surveillance for the 'monitoring, detection, identification, tracking, prevention, and interception' of irregular border crossings (Regulation, 2013). Surveillance creates a 'panopticon' across borders involving, apart from watching and observing, acts of processing, managing, and regulating individuals (Jumbert, 2012: 38), leading to a 'social sorting' logic (Lyon, 2003).

Regarding border control practices, information gathering, multilateral cooperation, technology, and professionalisation are the common practices extracted from the research. As a result, they compose the culture's border control practices.

On information gathering, border control is being conducted through data collection. This is being attested by EU official documents (European Commission, 2015a; Regulation, 2018), which instigate for a better use of information and its dissemination 'in a timely manner' (Regulation, 2016a, 2019). Irrefutably, this is why several computerised information-sharing mechanisms have been established, like EUROSUR (Regulation, 2013), ETIAS (Regulation, 2018), and the Entry/Exit System (Regulation, 2017). These systems collect, store, process, and analyse data, ensuring a smooth and swift data exchange. At the same time, by collecting data, they manage the production and dissemination of knowledge. So, they impact on the border control conduct by granting social control or, put differently, power to the actors that administer them (Jeandesboz, 2017), such as Frontex.

118 *Border control in process*

As far as multilateral cooperation is concerned, border control is being implemented practically through co-bordering (Longo, 2016). Migration is being handled as a matter of common concern that transcends national borders (European Commission, 2015a). Therefore, it requires a collaborative effort bringing on board, apart from member states, third countries. After all, Frontex joint operations are being conducted both within and outside the Schengen area with the participation of border officers from EU member states, Schengen states, such as Norway, and third countries, like Albania and Turkey. On that account, Frontex's multilateral cooperation with third countries shows that border control conduct goes beyond the framework set by EU institutions. An example of that is Frontex's cooperation with the UK before and after Brexit (analysed in Chapter 7).

On technological development, border control relies on the deployment of 'state-of-the-art technology' (Regulation, 2016a; Frontex, 2019a). Technologically mediated border checks, sophisticated monitoring devices, high-tech vehicles, and automated systems reconfigure borders as 'technological fortresses' (Marin, 2011) or 'cyber-fortresses' (Guild et al., 2008). Furthermore, technology enables borders' multiplication by expanding both the geographic border area and the time window for the border control conduct (Glouftsios, 2018). In fact, with ETIAS, the control of the intended entrant starts by completing the application for travel authorisation. Likewise, with EUROSUR Fusion Services, Frontex and national border control authorities can collect information about vessels suspected of illegal activities from satellites, thereby enlarging the monitoring area enabling for the tracking of the vessel of interest even before its departure from a third country's sea port.

The last border control practice is professionalisation. To cite the Schengen Borders Code, border control 'should be carried out in a professional [...] manner' (Regulation, 2016b). Likewise, high-level professionalism is being listed as a core value for the European Integrated Border Management (EIBM) (Frontex, 2019a). Over the last years, there has been an effort to professionalise border control through the development of specific skills and professional standards that foster a professional 'habitus' (Bigo, 2014) or the professionalisation of border guarding (Horii, 2012), which is distinct from other law-enforcement or military job categories (Olsthoorn & Schut, 2018). Fostering professionalisation, border guards or border control practitioners become a professional community united through shared preoccupations, demonstrating mutual support to their colleagues, and developing a common consciousness.

The aforementioned border (control) assumptions and practices constitute a border control culture schematically displayed in Table 6.1. Just browsing through them, it is possible to yield that these assumptions and practices reinforce each other. For instance, multilateral cooperation enables a smooth information gathering. Similarly, professionalisation and technology foster technocracy and vice versa, namely technocracy advances

Table 6.1 Warsaw border control culture

Border assumptions	EU external border
	National territory
Border control assumptions	Securitisation
	Technocracy
	Extra-territorialisation / Intra-territorialisation
	Surveillance
	Intelligence
Border control practices	Information gathering
	Multilateral cooperation
	Technology
	Professionalisation

professionalisation and technology. Moreover, intelligence-driven activities aim at a proactive action, which is promoted by a surveillance logic. At the same time, the performance of surveillance inside or outside the territory moves the border, accounting for extra- and intra-territorialisation, as well as merges internal and external security under a transborder securitisation context (Bigo, 2000).

And its name shall be Warsaw culture

The culture traced at Europe's borders rebuts Schengen as it does not share Schengen's border and border control logic. But why has Schengen been cast aside? Has there been any 'cataclysmic' event which triggered the formation of this culture? Do we have any new element at Europe's borders? Well, this new element and variation from Schengen can be no other than the advent of Frontex.

After Frontex's creation, border control evolved. New patterns of action, habits, structures, and social relations (Græger, 2016: 479), which were introduced or promoted by Frontex, were developed. These elements were not part of the Schengen regime. In parallel, ever since its establishment, Frontex grew, acquiring additional powers and tools to fulfil its expanding mandate. In other words, despite being an agency and therefore integrated within the EU institutional machinery, it started taking on a life of its own (Trondal & Jeppesen, 2008: 421). By developing its own vision, values, bests standards, working procedures, codes of conduct, glossary, model of risk analysis, knowledge, and social relations with other actors (Parkin, 2012: 1), Frontex progressively constructed and then fostered its own conception and implementation of border control.

One of the most important changes in relation to the pre-Frontex border control institutional regime concerns the integrated border management,[1] which has now become a shared responsibility between Frontex and member states (Regulation, 2016a, 2019). This responsibility sharing describes a different institutional reality as well as creates a variant legal framework regarding border control, with Frontex taking over new duties and responsibilities. In fact, to ensure the effective application of the EIBM at the external borders, Frontex transformed into the European Border and Coast Guard (Regulation, 2016a). The EIBM is considered as a key element for border management. Besides being implemented in the EU or Schengen area, the EIBM concept has been promoted even outside the EU, as attested by several EUBAM missions conducted — cooperating with Frontex — to foster non-EU countries' border security.

The above indicate that Frontex's impact on border control is not limited to the reshaping of the border control institutional architecture triggered by Frontex's establishment or changes in the everyday border control conduct produced by Frontex's activities at the borders. Most importantly, Frontex has reconstructed the EU border control by promoting a new border control culture variant of Schengen.

Some may have expected the opposite. That Frontex, being a product of the Schengen culture, would never have developed elements alternative to Schengen (Wolff & Schout, 2013; Ekelund, 2014). Yet, the examination of Frontex and borders identified evolved cultural elements. So, what remains to be seen is whether those elements constitute an alternative culture antagonising Schengen as well as how really these cultural elements differ from the other cultures of border control, which will be the focus of the next section. Disputing Frontex's impact, others may have argued that even if a new culture has emerged, that would be due to other actors and not Frontex, given that Frontex is just a dependent instrument or a vehicle and not an actor (Jorry, 2007; Colimberti, 2008; Vaughan-Williams, 2008; Léonard, 2010; Chillaud, 2012; Horii, 2012; Pallister-Wilkins, 2015; Paul, 2017; Csernatoni, 2018).

Nevertheless, Frontex's engagement with border control triggered certain significant changes in the border control conduct. To start with, a new border control actor, that is, Frontex, commenced undertaking operational activities at Europe's borders and even beyond. In doing so, Frontex deployed new border control tools. In fact, the research at the borders (Chapters 4 and 5) detected specific operational activities and tools that were imported at these two borders by Frontex, like dog and boat patrolling as well as thermo-vision cars in Evros; satellite monitoring and surveillance systems, including drones, in Lampedusa; debriefing on both borders. These constitute new elements for the border control conduct, highlighting an altered reality on the ground.

At the same time, since its establishment and operational launch, Frontex has become a member of the border control community, initiating therefore a change in the community's composition. By being part of it, it has reconstructed it. After all, a community cannot remain frozen. Rather, it adapts to shifts in the social environment, becoming reformed and reshaped (Adler, 2008; Kitchen, 2009). Due to its border control role, after

Border control in process 121

its establishment, Frontex started interacting with national border control authorities and especially national border guards. Actually, Frontex developed into a new 'colleague' for national border control authorities and a tangible symbol of the EU at the borders, bringing together all the national border control actors under its auspices and emblem. Thus, Frontex's addition in the border control community triggered a change. It altered the community's social environment and, in turn, the border control community. With Frontex's establishment and function, the border control community became restructured, putting border control practitioners at the forefront. For this reason, I label this community as 'practitiocratic', referring to the actors that practise border control, carrying out border control tasks and therefore being directly and actively involved with the border control conduct.

In addition, Frontex brought new border control assumptions and practices. Frontex, as a border control actor, is composed of certain assumptions and practices which reflect Frontex's spirit and method for border control. These are embedded in its rationale and action, as shown in Chapter 3. Therefore, Frontex, being a social actor and participating in a community, diffuses these cultural elements to the other members of the community through interaction, learning, and socialisation (Saurugger, 2013: 894, 2014: 152–154). Mutatis mutandis, not everything can be embodied by the other members of the community. Some things require more time for adaptation, whereas others may never be accepted and inserted in the community's rationale. As the research illustrated, there were certain elements not encountered at both the actor, Frontex (Chapter 3), and the border settings, Evros and Lampedusa (Chapters 4 and 5). For instance, the assumption of fundamental rights and the practice of policing were traced solely in Frontex's analysis but not at both borders. Therefore, they cannot be considered as constitutive parts of the assumptions and practices that form the border control culture. That said, there are common assumptions and practices encountered in both Frontex and at the borders. These compose the border control culture (see Table 5.1).

The above indicate that Frontex has a key role in this culture. This assessment is also shared by certain border control practitioners, namely Frontex staff and national border guards, as shown below:

> [Frontex officer] 'There is now a common culture […] Frontex promoted this culture'
>
> (Interviewee 1).
>
> [Frontex officer] 'Frontex is developing this culture'
>
> (Interviewee 3).
>
> [National border officer] 'Frontex has created a common culture'
>
> (Interviewee 12).

For this reason and to underscore Frontex's inclusion in EU border control as a new element at Europe's borders, I have chosen to label this culture as Warsaw, given that Frontex's seat is in Warsaw.

Schengen, Westphalia, Brussels, and Warsaw: Variant in name only?

The names Schengen, Westphalia, Brussels, and Warsaw compose 4 different 'loci' in geography. Aside from belonging in different European countries, they also represent variant approaches for border control. 'Westphalia', 'Schengen', and 'Brussels' constitute typologies of border control cultures in Europe developed by Ruben Zaiotti (2011). However, the current research at the borders and at Frontex revealed an alternative regime for border management, which this book decided to label 'Warsaw'. Comparing Zaiotti's border control typologies with the current border control conduct, it appears that Warsaw differs substantially (see Table 6.2). So, Warsaw accounts for a new border control culture.

In particular, fundamental differences emanate from the composition of the border control community in each border control culture. Westphalia is composed of officials from national governments and has a nationalist or governmental identity. Brussels portrays a supranational character manifested by EU officials and particularly officers from the European Commission. Schengen incorporates a regionalist or intergovernmental identity with a dual representation of officials from national governments and the EU. Warsaw's border control community, instead, can be characterised as 'practitiocratic' due to its border control practitioners' composition. The notable difference between Schengen and Warsaw is that Schengen's community comprises an elite of 'top' level officials, such as Interior Ministers or members of the Council of the European Union and the European Commission. Conversely, Warsaw's community is composed of the actual border control practitioners or the 'bottom' level, namely border guards and Frontex staff.

Regarding border assumptions, Westphalia considers borders as linear and barriers, stressing the need for stricter border controls so as to safeguard the territorial integrity and ethnic homogeneity of the sovereign state. Schengen perceives borders as semi-linear, functioning as filters by distinguishing internal and external borders and therefore internal and external cross-border flows. For Brussels, borders, being a symbol of Europe, function as bridges uniting people without any internal/external divisions (Zaiotti, 2011: 26). The above indicate that Westphalia's, Schengen's, and Brussels' underlying border assumptions do not match Warsaw's twofold conception of borders as national and EU external borders.

On border control assumptions, Westphalia perceives border control as governmental and national with a clear distinction between the internal and external field that nourishes a military emphasis (Zaiotti, 2011: 45–54). For Schengen, border control is trans-governmental, security-focused, and with

Table 6.2 Cultures of border control in Europe

	Westphalia	Schengen	Brussels	Warsaw
Period	1940s–1980s	1985–2000s	1985–1990s	2010s–Today
Borders	linear, barriers	semi-linear, internal / external distinction	bridges, no internal / external	twofold role: national border & EU external border
Border Control	national, governmental, strict, military emphasis	trans-governmental, asymmetric responsibility, security emphasis	supranational, balanced responsibility, economic emphasis	securitisation, technocracy, surveillance, intelligence, extra- / intra-territorialisation
Practices	unilateral, formal	trans-governmental, flexible	supranational, multilateral, legalistic	information gathering, multilateral cooperation, technology, professionalisation
Community	national (governmental)	regional (inter-governmental)	supranational (European)	practitiocratic
Community members	officials from national governments	officials from national governments & EU officials (Council, Commission)	EU officials (Commission)	border control practitioners (border guards & Frontex)
Reference texts	Montevideo Convention, UN Charter, National Constitutions	Schengen acquis	Single European Act, Maastricht Treaty	Frontex Regulations, EUROSUR Regulation, EIBM Strategy

Note: The elements included in Westphalia, Schengen, and Brussels cultures are based on Zaiotti's typology (2011: 26).

an asymmetric distribution of responsibility among the EU states (Zaiotti, 2011: 71–74). Following Schengen's spirit, there is free circulation in the Schengen area, with a parallel hardening of security provisions at the external border. Brussels, whilst prioritising the economic dimension, considers border control as supranational, advancing an approach of balanced distribution of responsibility for border control among member states (Zaiotti, 2011: 80–84). For Warsaw culture, border control is based on securitisation, technocracy, extra-territorialisation/intra-territorialisation, surveillance, and intelligence. The short overview of border control assumptions shows that although technocracy, surveillance, and intelligence are constitutive elements of Warsaw culture, they are missing from the other border control paradigms. Differences are manifested even among shared elements. In particular, both Schengen and Warsaw include security; Schengen's focus, though, is on national security, while Warsaw's on securitisation. For Warsaw, securitisation is a social process (Williams, 2003: 523) or a pursuit (Buzan, 1991: 37). Actually, securitisation refers to a process that frames an issue as a security concern. Being securitised, this issue becomes perceived and handled as an existential threat requiring the adoption of emergency measures and the suspension of 'normal' politics in dealing with it (Buzan et al., 1998). For Schengen, however, security constitutes a normative state. Following that, security preoccupations constitute matters of national security. Therefore, they are dealt with the standard political system by the central government. Likewise, on the geographic dimension of borders, both Westphalia and Schengen refer to an internal/external territorial distinction. Yet, for Warsaw this is described as extra-territorialisation and intra-territorialisation, marking a shift in the space of border control. This shift is continuous as opposed to a static internal/external distinction. Actually, the extra- and intra prefixes denote a deeper and wider context than Schengen's internal/external frames.

Turning to border control practices, Westphalia is characterised by formal and either unilateral or bilateral practices, with borders being clearly marked by the sovereign state so as to retain control over its territory, and has a spirit of protectionism, to limit imported goods based on unilateral assessment or in line with bilateral agreements (Philpott, 2001: 12–13; Zaiotti, 2011: 26). Schengen is organised around trans-governmental and flexible border control practices materialised through the adoption of ad hoc arrangements for enhanced cooperation that enable member states to pursue deeper integration. Brussels' practices are supranational, multilateral, and legalistic (Zaiotti, 2011: 26). Accordingly, the Brussels model evokes the abolition of all border controls, implemented through the delegation of power to a supranational authority. This would result in the enhancement of the judicial oversight of the border control policy-making process and therefore its democratic accountability (Zaiotti, 2011: 233). Warsaw's practices include information gathering, multilateral cooperation, technology, and professionalisation. So, Warsaw's border control practices enclose a different spirit than those of Westphalia, Schengen, and Brussels. Actually, Warsaw's are more concrete and refer to the border control conduct instead of the border control policy, manifesting therefore the shift from policy-makers to the border control practitioners pursuing Warsaw culture and composing Warsaw's community. Apart from multilateralism, no other element of Warsaw is encountered as a practice in Schengen, Brussels, or Westphalia. Even multilateralism is conveyed and materialised differently in Brussels and Warsaw. In Warsaw, multilateral cooperation describes a multi-actor setting with the deployment of border guards and equipment from other member states as well as the construction of formal and even informal collaboration channels with various stakeholders. Conversely, Brussels' multilateralism refers to formal cooperation framed by legal provisions.

The variance of these cultures is also illustrated by the separate texts that embody their spirit. Westphalia culture is internalised in historical documents, such as the 1933 Montevideo Convention (Albahary, 2010). Schengen's common sense is being presented in the Schengen acquis (Walters, 2002; Marenin, 2010). Brussels, instead, has been inspired by the Single European Act and the Maastricht Treaty (Callovi, 1992; Vatta, 2017). All these documents represent a different border control mentality, given that they were configured in separate historical periods. Westphalia was a dominant culture for border control in Europe until the 1980s. Schengen, which replaced Westphalia, became the new dominant culture of border control in Europe in the 1990s, whereas Brussels, which was also a product of the 1990s, has never reached its full maturity and thereby did not succeed in becoming a dominant culture.

The chronological period during which these cultures were formed and the documents upon which they were based seem to have become outdated, as they were developed to respond to the needs and preoccupations of bygone circumstances. Accordingly, Schengen culture, which was the last dominant culture for border control, reproduces the assumptions and practices for border control integrated in the Schengen Agreement, that is, a border control spirit and reality that was relevant almost 40 years ago. Today, certain aspects of this reality, if not all, seem obsolete. Borders have changed. As such, the tools and actors involved in border control have altered. Likewise, Europe is not the same. After fundamental crises, membership expansion — even membership deduction — deepened integration, and the transfer of new competences in the field of border management, a new context has emerged that Warsaw culture seeks to address. Traces of this culture can be found in contemporary texts, like the EUROSUR Regulation (Regulation, 2013), the last Frontex Regulations (Regulation, 2016a, 2019), the European Agenda on Migration (European Commission, 2015a), the Technical and Operational Strategy for the European Integrated Border Management (EIBM Strategy) (Frontex, 2019a), the Schengen Borders Code (Regulation, 2016b), the ETIAS Regulation (Regulation, 2018), and the Smart Borders Package, which includes the Entry/Exit System (Regulation, 2017). So,

Warsaw culture is a product of today's environment. It outlines the current border control regime. It refers to the present, not to the past.

The description of this new culture, manifests Frontex's key role. Actually, the establishment of Frontex constitutes a variation from the composition of the Schengen border control community, as it did not exist during the consultations and development of the Schengen paradigm. Nevertheless, as shown in this research, after its creation, it became a significant border control actor and member of the border control community, ascribing to it the characteristic 'practitiocratic'. In parallel, by constructing new intersubjective meanings (Haas & Haas, 2002), such as underlying assumptions and practices, Frontex shaped the border control culture. So, naming this culture 'Warsaw' is not that daring or surprising. Having analysed the composition and characteristics of this culture, the next step is to investigate the move from Schengen to Warsaw culture.

From Schengen to Warsaw

According to Zaiotti (2011: 27–43), the transition from one culture to another occurs through a process of cultural evolution. Through this process we can trace how Warsaw culture became the new 'norm' in EU border control, taking the place of Schengen.

As already presented in detail in Chapter 1, the first step for a cultural evolution is culture's variation. Cultural variation refers to the emergence of an alternative border control culture (Zaiotti, 2011: 31–32). Indeed, the previous section demonstrated that Warsaw culture varies substantially from the other cultures of border control, namely Schengen, Westphalia, and Brussels. This illustrates that a cultural variation has already occurred. Indeed, the emergence of new assumptions and practices with Warsaw challenged the pre-existing Schengen cultural elements. So, the dominant culture, Schengen, has become challenged. The members of the border control community, instead of moving towards the already formed alternative cultures, namely Brussels or even Westphalia, concocted a new culture, that is, Warsaw. The creation of Frontex was an essential and critical juncture that visibly shifted the circumstances at EU borders and activated this cultural variation.

Frontex's launch and its engagement with border control created a new reality at the EU borders that could not be explained by Schengen's tenets formed in the 1990s. Still, Frontex was not the sole change in relation to Schengen's regime. New technologies, more intense migratory pressures, increased cross-border mobility represent just a few of the last years' developments affecting borders and border control in Europe (Tholen, 2010). All these shifts, unavoidably, triggered questions about Schengen's relevancy (Carrera et al., 2013), leading, in turn, to Schengen's variation.

The next step in the cultural evolution process is culture's selection. It involves two phases. Firstly, culture's pursuit; secondly, culture's anchoring

(Zaiotti, 2011: 33–37). The pursuit of Warsaw culture started after the development of the 'practitiocratic' community. As already mentioned, the community of Schengen culture was composed of Ministers of Interior and 'top' EU officials from the European Commission or the Council. During Schengen's consolidation, border control practitioners did not have many chances to communicate and interact with their European colleagues. Despite being the persons that were actually conducting border control, they were left on the fringes of their national borders. In fact, everything about border control was decided away from the borders. That changed when border control practitioners realised their common disposition. After all, border control practitioners 'know much more about borders and what Europe is than politicians' (Interviewee 2). The establishment of specific work standards, rules, training curricula, glossaries, and codes of conduct harmonised border control actions (Paul, 2017), diffusing a spirit of 'borderguardship' (Frontex, 2015b: 109). Similarly, the initiation of certain socialisation activities, like joint operations, exchange programmes, field visits, workshops, and seminars, enabled border guards to communicate, exchange views, form a connection, and start perceiving border guards from other member states as their colleagues or even friends (Interviewee 2 and 9). The same applies to challenges at the borders or difficulties in border control, as, by border guards participating in joint operations away from their home country, they became common preoccupations for the border control practitioners (Interviewee 1). All these diffused a common sense among border guards and progressively built the feeling of a community. In this light, border control became a 'joint enterprise' (Adler, 2008: 199). This community, however, could not be related to the Schengen culture, as border control practitioners were not part of it. They were not included in its design and development. As a result, border control practitioners started forming and pursuing a different approach that seemed more relevant to their working environment and more effective in addressing contemporary border control preoccupations. So, the pursuit of a new culture stemmed from a reasonable decision of the members of the community to address border control challenges efficiently (Zaiotti, 2011: 34).

Subsequent to the decision to pursue an alternative culture is its anchoring, which includes culture's performance testing and, if deemed successful, its collective adoption by the members of the community (Zaiotti, 2011: 34–37). Warsaw's testing occurred through Frontex's operational activities and especially its joint operations, as they incorporated the assumptions and practices of Warsaw, testing therefore their materialisation. For instance, the operational presence of Frontex at the borders reflected the assumption that, besides being national territories, European borders also enclose an EU dimension. Moreover, the method upon which these joint operations are based, such as risk analysis, multilateral cooperation, information gathering, the use of technology, incorporates Warsaw's assumptions and practices, enabling their experimentation. 'Hard' proof assessing

Warsaw culture's performance is constituted by the data collected regarding the apprehended irregular border crossers, overstayers, smuggled goods, returned migrants, and other cross-border crimes, which Frontex regularly publishes as a means to assess its operations. The outcome of this evaluation has been judged positive, leading to Warsaw's collective adoption. This is being attested, for instance, by Frontex's continuous mandate enhancements as well as the rise in Frontex's budget and the number of its joint operations, as presented previously (see Chapter 2). The same conclusion can be drawn from the first RABIT deployment. When Greek authorities requested in 2010 the activation of the RABIT mechanism during extreme migratory pressure at the Greek-Turkish land border, they deemed both Frontex and the border control model that it promoted to be effective solutions to tackle a border crisis. Apart from member states, this border control regime implemented by Frontex became increasingly accepted among third countries. For this reason, in 2009, just 4 years after the first joint operation, border guards from neighbouring third countries, like Russia, Ukraine, and Belarus, started participating in Frontex joint operations (Frontex, 2010b: 35). All these indicate that Warsaw's culture testing has been considered successful, leading therefore to its final adoption by the border control community.

The final step for a cultural evolution is culture's retention. It refers to culture's institutionalisation, which takes place with the integration of its assumptions and practices in a legally binding context (Zaiotti, 2011: 42). In the last decade, there has been a significant expansion of border control documents, which formally incorporate the elements of Warsaw culture into the border control policy domain. For instance, the 2013 EUROSUR Regulation fosters surveillance by setting up a sophisticated surveillance system, whereas the 2016 Frontex Regulation turns border management into a shared responsibility, reflecting the dual nature of EU borders. In addition, the EIBM Strategy encompasses technology (Frontex, 2019a). The most striking is that the word 'technology' was missing from earliest definitions of integrated border management (Council of the EU, 2006a). The adopted EU documents manifest a change in the border control rhetoric and a diverse border control approach accompanied by new measures and tools. Besides confirming this change, these documents function as a channel for the institutionalisation of the new culture's tenets, leading therefore to Warsaw's retention.

So, this section showed that a cultural evolution has already occurred. All the steps and phases in Zaiotti's cultural evolution mechanism have been traced here in the same chronological sequence. Such sequence evidence (Beach & Pedersen, 2013: 178) points out that Warsaw constitutes a new culture for border control in Europe. Better said, since the finalisation of the cultural evolution with the last phase of cultural retention, Warsaw now constitutes the dominant culture of EU border control, replacing Schengen.

Frontex in Warsaw

Why is the emergence of a new border control culture of any relevance to Frontex? Why link Frontex to Warsaw, aside from being the geographic location of its seat? Here, I am assessing Frontex's role in the development of this culture, breaking down its impact on the formation of Warsaw's constitutive elements. A border control culture is composed of a set of border control assumptions and practices. To become shaped and materialised this culture depends on a border control community and, concomitantly, on reference texts that institutionalise its main tenets. These elements, apart from being cultural components, also constitute essential conditions for a border control culture to develop. Therefore, to explore Frontex's role in the consolidation and promotion of Warsaw border control culture, the impact of Frontex on these three conditions is being assessed. The investigation follows a process-tracing research design that includes the formulation of predictions so as to trace Frontex's impact (George & Bennett, 2005: 206).

The condition of the border control community

To validate that Frontex is promoting Warsaw culture, I formulate the prediction that Frontex has developed Warsaw's border control community. Empirical evidence confirms this prediction: Frontex as a border control actor enabled and, at the same time, formed Warsaw's 'practitiocratic' border control community. Thanks to Frontex, border control practitioners came into direct contact to construct a community.

The existence of this community has been empirically manifested by the analysis in Lampedusa and Evros (Chapters 4 and 5). National border guards work shoulder-to-shoulder at the border with Frontex guest officers and officers from Frontex's standing corps, interacting as members of the same community. They share their experiences, discuss work challenges, confront similar problems, and even set common professional goals, establishing a connection with their colleagues

Now, regarding Frontex's contribution, sequence evidence attests Frontex's role in its development. The institutional structures that existed before Frontex, like SCIFA+, referred solely to the heads of member states' border control services; hence, top-level officials, most of which did not stem from border guarding services. Rather, they belonged to a broader professional corpus, such as the Police, who were meeting to conduct preparatory discussions about migration, border control, and asylum issues for the Justice and Home Affairs meetings of the Council of the European Union. So, before Frontex there was no opportunity for border control practitioners from different EU member states to meet and discuss practical matters of common interest. This irrevocably changed with Frontex. Frontex joint operations enabled border officers that were not high-ranking staff to go abroad, guard borders other than their national, and start working shoulder-to-shoulder with border guards from different countries. Besides joint operations, additional chances of interaction emerged during border guards' participation in various activities organised mostly by Frontex, such as workshops, conferences, and training seminars. Flagship among these events is the EBCG Day initiated in 2010 by Frontex. Since then, it has taken place annually in Poland with the aim to strengthen 'Europe's border guard community' (EBCG Day, 2022). Fostering this community, Frontex has even developed a European Joint Master's Programme in Strategic Border Management, which constitutes the highest education accreditation for border guards. All these indicate that for border guards:

> [National border guard] 'Frontex has broadened our horizons'
> (Interviewee 5).

The harmonisation of border control tools and policies drives towards the progressive development of a common sense among border control practitioners. To this end, Frontex produces handbooks, codes of conduct, reference material, technical programmes, specialised training, and education; as well, it develops common indicators, for instance for risk analysis or evaluation standards. All these nourish shared understandings and common routines guiding, in turn, actions and behaviour. Border officers start working in a similar way (Interviewee 12). Simultaneously, shared understandings and common routines function as a transmission channel to other actors (Interviewee 2), weaving together border guards. Hence, through communication and socialisation they find more things uniting them than keeping them apart. So, a 'we feeling' starts taking root (Nathan, 2006: 276). In other words, apart from developing it in its early stages, Frontex also sustains this community by nourishing and maintaining a professional 'habitus' (Bigo, 2014) or an esprit de corps among border guards (Frontex, 2014: 49). For instance, with each Frontex publication, such as risk analysis reports, glossaries, and handbooks, Frontex conveys and transmits the community's language or, better, its meaning of discourse (Crawford, 2002: 65). In this spirit, it has also produced a monthly newspaper labelled 'The Border Post' that keeps the border guard community informed about current border control trends and Frontex's actions.

Turning to account evidence, the first phrase of this chapter shows that Frontex acknowledges the existence of this community, as its Executive Director openly talks about border guards' 'community'. Similarly, a community spirit and like-mindedness is reflected in several narratives from border guards and Frontex staff, like:

> [Belgian officer] 'We are constantly learning from each other. We are building a network of officers that you can always ask for help'
> (Frontex, 2018).

[German officer] 'Guest officers are like a large family now'
(Frontex, 2014: 27).

[Head of Analysis and Planning Sector] 'One of the good things about the EU border guard community is that they're open and keen to work together'
(Frontex, 2010a: 65).

So, Frontex gives to the members of the community not only the space and occasion for interaction, but also the tools to interact and interpret the others, such as with the articulation of glossaries or common values presented in Frontex's codes of conduct. Furthermore, it produces specific signification signs (Wedeen, 2002: 720), which allow the community's identification through symbols, like the Frontex logo pictured in Frontex uniforms. All these constitute trace evidence of Frontex's role within the community.

In sum, by initiating these activities, Frontex contributed to the development of this border control community. It became both its initiator and active enabler functioning as a 'melting pot' (Interviewee 2). It could be assumed therefore that without Frontex the 'practitiocratic' border guard community could not have been developed and, in turn, the Warsaw culture could not have emerged. This reveals a causal role for Frontex.

[Frontex officer] 'Now we have a community [...] Frontex builds this border control community. Things were different before Frontex. They changed after Frontex'
(Interviewee 2).

The condition of reference texts

The next prediction is that Frontex has participated in the production of Warsaw's reference texts. This is another correct prediction as shown below.

Key reference texts for the Warsaw culture analysed here are the 2016 Frontex Regulation, the 2013 EUROSUR Regulation, and the 2019 EIBM Strategy. I have chosen these texts because they focus on variant border control aspects or developments, that is, surveillance, the European Border and Coast Guard, and the European Integrated Border Management. Furthermore, being tabled not simultaneously, but once in every three years, these texts have been drafted in the 2010s, namely after Frontex's formative years and when Warsaw culture became institutionalised. At the same time, they have been produced by different institutional actors. These differences enable to assess Frontex's role, recognising its impact in the most different Warsaw's reference texts.

Starting chronologically, the 2013 EUROSUR Regulation established the European Border Surveillance System, namely EUROSUR. According to Frontex's Executive Director, EUROSUR revolutionises EU border control by providing a 'pan-European dimension to situational awareness' (Frontex, 2012: 5). Crucially, with the employment of modern technological tools, it sets the European frontiers under a state of constant surveillance (Jeandesboz, 2017: 256).

Tracing EUROSUR's path, it was first proposed by the Commission (2006: 3). However, the Commission's proposal was based on two different studies prepared by Frontex, namely MEDSEA and BORTEC (Commission, 2008).

MEDSEA was delivered by Frontex in July 2006, that is, 4 months prior to the Commission's proposal. As a member of Frontex staff mentioned, 'the MEDSEA study was the foundation of everything which followed [...] [BORTEC and EUROSUR] came from MEDSEA' (Frontex, 2010a: 43). Regarding its content, MEDSEA constitutes a feasibility study for the development of a Mediterranean Coastal Patrols Network. Yet, it also addresses surveillance. In this vein, it states that 'surveillance [...] has to cover not just an entry point, but a variable-depth surface' (Council of the EU, 2006b: 11). This conforms to a wider monitoring in the pre-frontier area and not just at the border crossing point, which started to be materialised via EUROSUR and its later Fusion Services.

BORTEC was carried out by Frontex at the end of 2006 and was presented in January 2007. Though only a few parts of this study are publicly accessible, BORTEC constitutes a key document for the establishment of EUROSUR, because it explores the technical feasibility for its establishment (Wolff, 2012: 144). In particular, it presents the basic axes of EUROSUR's design, putting forward a 'system-of-systems' approach for border surveillance (Jeandesboz, 2017: 174), which places Frontex at its centre with the agency acting as its chief coordinator.

Both studies constituted the basis for EUROSUR's implementation, taking into account that Commission's initial proposal did not entail any concrete measure about EUROSUR's establishment or its components (Commission, 2006). Consequently, Frontex's studies became the reference guide for the articulation of both the EUROSUR proposal and the final EUROSUR Regulation. After all, the structure model proposed by MEDSEA, involving the creation of National Coordination Centres, and BORTEC's proposal for attributing to Frontex a coordination role are both included in the adopted EUROSUR Regulation. Thus, evidently, Frontex participated in the production of the EUROSUR text. Most importantly, the agency developed EUROSUR by both designing and operating it.

Another important text is the 2016 Frontex Regulation, as it transforms Frontex into the European Border and Coast Guard agency and stipulates a shared responsibility for the management of external borders between Frontex and member states. This document was drafted with Frontex's vivid involvement. In particular, the Regulation has been based on a proposal of the European Commission (2015b). The Commission's proposal, though, has been elaborated in line with the recommendations of Frontex's management board. In a straightforward manner, the Commission acknowledges that

Border control in process 133

the proposal 'reflects the majority of recommendations' of the management board regarding Frontex's enhancement (European Commission, 2015b: 7) clearly approving, in its entirety, Frontex's management board rationale. Apart from the input of Frontex's management board, the Commission's proposal and, in turn, the 2016 Frontex Regulation, have been tabled on the basis of feedback and suggestions provided by Frontex staff.

In fact, the Commission's proposal has been written in accordance to a feasibility study produced by the company Unisys (2014). Unisys' study was examining the prospect of the creation of a European System of Border Guards. Unisys conducted interviews with representatives of Frontex as well as visited Frontex's headquarters in Warsaw to collect the agency's expertise (Unisys, 2014: 11). 14 Frontex officers contributed information to this study, whereas regarding representatives of other stakeholders at EU level, only 7 Members of the European Parliament and just one officer from the European Commission were contacted (Unisys, 2014: 40–41). The difference in the number of contributors highlights the increasing importance attributed to Frontex's border control expertise as well as Frontex's ability in shaping the content of this document. Accordingly, several Frontex proposals have been included in the final text, like Frontex's strengthened role in the field of returns, the development of a common integrated risk analysis model, the establishment of common training standards, co-ownership with member states of technical equipment, and new responsibilities during emergencies (Unisys, 2014: 18–19; Regulation, 2016a).

The last document for investigation is the EIBM Strategy produced by Frontex in 2019. This strategy, though operational and technical, constitutes a key component of the EIBM at both national and EU level. It sets the parameters for the mapping and then implementation of an overall EIBM Strategy. Indeed, the 2016 Regulation states that any national Strategy for Integrated Border Management must take into account the EIBM Strategy drafted by Frontex. This means that Frontex, besides being responsible for the content of the technical strategy developed, also affects the national strategies of its member states, because they must be in accordance to Frontex's strategic document. After all, Frontex's EIBM Strategy chronologically precedes national strategies.[2] So, Frontex has already taken the lead. Actually, Frontex's Executive Director had in 2014 already stressed the necessity for the articulation of a renewed Integrated Border Management (EBCG Day, 2014). Frontex not only pushed for its reformulation but also defined its final trajectory. The EIBM Strategy produced by Frontex, even though it is considered as technical and operational, still constitutes a strategic document that entails a vision, values, priorities, strategic objectives, and proposed actions (Frontex, 2019a). These elements reset the whole border control policy and place Frontex at its core by rendering the agency a 'guardian' of the European Integrated Border Management (Frontex, 2019a: 17). Crucially, this Strategy reflects Frontex's vision for border control and the EU borders incorporating Frontex's and Warsaw's assumptions and practices.

Though this Strategy has been drafted after consultations with national and EU stakeholders, it still constitutes a text issued by Frontex. This is evident throughout the document; the Strategy's layout, terminology, structure, and even selection of colouring, all match with Frontex's publications. Thus, the EIBM Strategy constitutes a Frontex document, because it has been produced by Frontex and entails the agency's spirit. It should be noted that the agency is currently in a process of reviewing the EIBM document, and for this reason it has decided to establish a European Integrated Border Management Working Group. This Working Group will meet at Frontex's headquarters, whereas, according to Frontex's management board decision of 7 June 2022, the agency will provide it with 'the necessary substantive input'.

The above validate that Frontex has participated in the production of Warsaw culture's reference texts. As it was demonstrated by sequence, account, and trace evidence, Frontex had a key role in the drafting of the 2013 EUROSUR Regulation, the 2016 Frontex Regulation, and the 2019 EIBM Strategy. Frontex's role was not constrained to technical expertise or statistical data provided by the agency; Frontex was actively involved in tabling these documents, even shaping their content through the inclusion of Frontex's suggestions.

The condition of assumptions and practices

The last prediction is that Frontex has developed Warsaw culture's assumptions and practices. This is another correct prediction, given that Frontex has developed most of these cultural traits by converting them from border control elements into assumptions and practices for border control.

Starting with border assumptions, the consideration that borders constitute part of the national territory, of course, cannot be considered a Frontex conception. Rather, it is traced back in the Westphalian state-centric system (Zaiotti, 2011: 48). That cannot be the case for the assumption that borders are also EU external borders. Both the Treaty of Amsterdam and the Schengen Agreement introduced the term 'external border' at the EU lexicon. Progressively the EU started acquiring competence in the policy area of migration and border control, adopting new legal instruments and policy initiatives. Nevertheless, it is with Frontex that the 'EU external border' has been moulded from a policy issue or vocabulary term into a border assumption. Border guards do not perceive any more borders as being just their national borders but the borders of Europe (Interviewee 1). Prior to Frontex, national officers were managing their national borders, which, institutionally and legally, were also functioning as EU external borders. After Frontex's establishment and the deployment of border officers from other EU member states, these borders truly functioned as EU external borders

Border control in process 135

signalling the prevalence of a new border assumption. Characteristic is the following narrative of a border officer deployed to Greece:

> [Slovenian officer] 'I'm European, and I consider Greece's borders to be our borders too'
>
> (Frontex, 2019b).

Turning to border control assumptions, certain pre-existing elements encountered at the borders were transformed into shared border control assumptions after Frontex's establishment and due to Frontex's active contribution. Others that were not part of the EU border control before Frontex were initiated and diffused by Frontex, evolving progressively into border control assumptions.

Over the last decades the view that border control is diffusing a securitisation logic has become quite common (Huysmans, 2000). But since Frontex, this securitisation context from a general notion and inclination became formalised with concrete terminology and working methods, like the CIRAM tabled to assess possible border control risks.

By contrast, technocracy was never a border control characteristic encountered in every EU external border (Baldwin-Edwards & Fakiolas, 1998). It remained dependent on national choices for policy design and implementation. The transition to a technocratic conception of border control was greatly accelerated with the creation and function of Frontex. Frontex, being an EU agency, was established as a technocratic actor (Parkin, 2012: 1). Moreover, Frontex's very function promoted technocracy. By obliging member states to conform to its border control methodology through the preparation of operational plans for its joint missions or the development of new computer platforms for data uploading and sharing, like JORA, Frontex actually spread a technocratic mentality across the EU borders or, as Campesi put it, a 'technocratic utopia' (2022: 162).

Regarding extra-territorialisation and intra-territorialisation, the shifting of the border control 'topos' has been a typical characteristic of the meta-Westphalian era (Agnew, 1994). However, Frontex both intensified and normalised this process. Shifting border management away from the fixed geographic borderline became formalised and routinised, for instance with surveillance in the pre-frontier area implemented via EUROSUR. As a Frontex officer has mentioned, in 5 years from now there will be no concrete geographic point functioning as a border. Instead, border control will be conducted through the collection of background information about indented travellers (Interviewee 3), like the ETIAS logic. So, border becomes less and less a geographic point on a map.

Intelligence constitutes another assumption, and whilst not coined as a term by Frontex, this agency has rendered it a central border control element by promoting an intelligence-driven border control. In fact, Frontex is considered an intelligence-led agency (Wilson, 2018: 46). It has integrated intelligence in most of its border control functions. For instance, every Frontex operation is intelligence-driven. Furthermore, the agency has established intelligence communities to collect data and intelligence sharing platforms, like FLO and TU-RAN. All these measures have both operationalised intelligence and rendered it an EU border control assumption.

The last border control characteristic is surveillance. The empirical research at the borders has already manifested that Frontex installed modern surveillance systems in both Evros and Lampedusa. Moreover, the previous part uncovered that EUROSUR, namely the system that for the first time offered real-time pan-European surveillance, has been developed by Frontex studies. Hence, though Frontex did not invent the wheel of surveillance, it enabled its implementation in an EU context by putting EU borders under constant and total surveillance (Jeandesboz, 2017: 256). So, Frontex vastly increased the scale of conducted surveillance as well as the degree by which border control became reliant on surveillance.

Likewise, Warsaw's border control practices were not part of a cultural trajectory before Frontex. Nevertheless, they turned into border control practices via Frontex's function. Heightening and integrating them in its border control activities, Frontex enabled their development as Warsaw's border control practices.

Starting with information gathering, this is one of the reasons that Frontex was created in the first place, that is, to enable and ensure information dissemination (Regulation, 2004). Information gathering was always a border control preoccupation. Still, it was not an achieved goal. Frontex, responding to this need, started functioning as an information hub. It built new IT reporting systems, like JORA and EUROSUR, which allowed information gathering and sharing. In addition, to facilitate information management, Frontex proposed the establishment of new permanent structures in each member state at both local and national level. As a result, National Coordination Centres (NCC) and Local Coordination Centres (LCC) were created (Frontex, 2022).

Multilateral cooperation and Frontex could be perceived as synonymous. Certainly, multilateral cooperation was not initiated by Frontex. Instead, it can be traced back to the Schengen initiative as well as the formation of certain bilateral or regional cross-border cooperation structures.[3] Still, Frontex's functions transformed multilateral cooperation from a contractual provision or ad hoc institutional setting into a recurring organised activity, namely a border control practice. After all, it cannot be random that Frontex missions are named joint operations. This name underscores that Frontex operations cannot be conducted solely by one member state but only jointly.

As for professionalisation, before Frontex the pursuit of professionalism in the field of border control did not constitute an EU border control priority. Instead, it rested upon national systems to promote their border control authorities' professionalisation, leading as a result to severe differentiations across Europe. This omission at the EU level is reflected by the fact that the European Commission's Communication for Integrated Border Management did not include any reference to professionalism (Commission, 2002). Since Frontex, this has changed. The agency has made professional excellence its 'institutional mantra' (Frontex, 2014: 15) compelling it to urge for professionalism. Also, Frontex has listed professionalism as the first value in the EIBM Strategy (Frontex, 2019a: 11). All these show that Frontex has enabled professionalisation to become an EU border control practice.

Last, but not least, is technology. The advancement of technology has been accompanied by its proliferation in the border control conduct. This constitutes a global trend that transcends EU borders. Trying to keep up with this development, the European Commission has repeatedly prompted its member states to employ new technologies for border checks (Commission, 2002). Yet, this is another issue that rested upon the national level, being dependent on budget restrictions and policy choices. Frontex tried to remedy this. For example, it has brought modern technological tools to the EU borders during its operations, such as thermo-vision vehicles and aerostats at the Greek-Turkish border. In addition, the agency tests and proposes new technological solutions for a harmonisation of technological capacities across the EU and innovative border management products. In this context, since its creation, Frontex has not only promoted but also largely insisted upon the use of biometrics and automatic border crossing (automated border control — ABC) systems (Interviewee 3), producing studies, operational guidelines, and best practices as well as organising conferences. Measuring Frontex's impact, in 2011, 7 member states were operating or testing ABC systems (Frontex, 2011), whereas in 2015 there were 14 (Frontex, 2015c). This constitutes pattern evidence of Frontex's impact. Up to a certain extent, one can link this significant increase to the general technological progress. Yet, to be fair, it can also be attributed to Frontex's fixation with these operative solutions. Actually, while the use of technological innovation is expected to take place in some states, the application of technological solutions by states that are notorious for their ineffective bureaucracies or non-well-equipped policy authorities is not a foregone conclusion. So, Frontex has enabled and intensified the use of technology for border control purposes.

The analysis, drawn from trace, pattern, sequence, and account evidence (Beach & Pedersen, 2013), validates the prediction that Frontex has developed (almost all) Warsaw culture's assumptions and practices; with one exception, the first border assumption, namely that borders are part of the national territory, which derives from the Westphalian regime. All the other assumptions and practices for borders and border control were transformed into shared cultural traits due to Frontex's contribution. In sum, with Frontex's help these border control elements, from a fragmented adoption dependent on national prioritisation or an EU aspiration limited to text references, were converted into border control assumptions and practices redefining, in turn, the border control conduct at Europe's borders.

138 *Border control in process*

Having assessed Frontex's contribution, the next section moves to the EU institutions, discussing their input into the emergence and development of the Warsaw culture. Bringing the EU institutions into Warsaw's analysis allows getting a clearer picture about Frontex's impact and its possible limits or liberties that form its power and causal role in Warsaw's culture of border control.

Out of Frontex's box: EU institutions in Warsaw

The relationship of EU institutions with Poland in the last years may mildly be described as strained, especially after the recent rule of law battle between Brussels and Warsaw (European Parliament, 2021), triggering even discussions about a 'Polexit', namely the possible exit of Poland from the EU. Apart from using the word 'Warsaw' to refer to Poland's authorities or government, this book coined and proposed the term 'Warsaw culture', that is, a culture of border control pursued at Europe's borders with distinct assumptions, practices, and community members. Focusing on Warsaw as a border control culture, I will now discuss EU institutions' stance towards Warsaw (culture).

Before Frontex, when Warsaw culture did not exist, the EU institutions were promoting and pursuing Schengen, as it was the dominant culture for border control in Europe. Actually, according to Zaiotti's analysis (2011: 26), Schengen's community composition included two different levels: the national and the EU. The national level, which had the lion's share due to Schengen's intergovernmental nature, included officials from national Ministries, mainly Ministries of the Interior. At the EU level, Schengen's community involved officials from the Council and the Commission. These EU officials, challenging the nationalist approach of Westphalia that was the then-dominant border control culture, both elaborated and pursued the Schengen regime, achieving its subsequent ascendancy through a cultural evolution process (Zaiotti, 2011). So, both the Council and the Commission were actively pursuing Schengen culture as they were part of its border control community. After all, even their decision, or compromise, to create Frontex was integrated into the Schengen-led rationale, with Schengen being at the time the unchallenged and still-dominant culture of border control in Europe (Zaiotti, 2011: 168).

Turning to the European Parliament, the most democratic EU institution had no role in Schengen's formation (Zaiotti, 2011: 74). Rather, it accepted it when the Schengen trajectory became finalised and approved by the members of the border control policy community, in which it did not participate. The European Parliament's assent to Schengen was expressed, in a rather vivid manner, when the Parliament started expressing its approval for texts that reflected Schengen's spirit, like the 1997 Treaty of Amsterdam. By recommending the Treaty's ratification, the Parliament endorsed a text that incorporated the Schengen acquis into the EU framework.

Being part of the Schengen culture (Council and Commission) or just endorsing it (Parliament), the EU institutions did not take part in Schengen's contestation. Rather, Schengen became challenged when the practitiocratic border control community, developed by Frontex, started the pursuit of Warsaw's cultural trajectory. Although they were not part of the Warsaw community, the EU institutions became accomplices in Schengen's demise. Establishing a new institutional structure, that is, Frontex, and calling for more effective management of the EU external borders (Commission, 2003), they were recognising that Schengen regime needed enhancement or different tools to address more efficiently emerging challenges at the EU external borders. Thus, they inadvertently paved the way for Schengen's contestation. Similarly, by creating a technocratic agency that collected all the experience and specialised knowledge on such a technical field, they gave Frontex a comparative advantage in relation to the EU institutions that were not actively involved with border control, such as the Parliament or the Commission. So, the EU institutions contributed to Warsaw's formation and then to Schengen's demise. Nevertheless, that was an unintended consequence and not a conscious choice from their part.

By contrast, their participation in Warsaw's retention constituted a deliberate decision. After approving the operational success of this new culture, the EU institutions embraced Warsaw's cultural trajectory by both allowing and contributing to its institutionalisation with the integration of its key tenets in official documents and fencing it in a legally binding context. The change in EU institutions' stance and their adoption or even promotion of Warsaw's principles can be identified by a shift in the content of relevant border control EU documents. For instance, the 2004 Council Regulation establishing Frontex did not include any reference to a 'shared responsibility'. After all, such a provision transcends Schengen's intergovernmental cultural trajectory. The 'shared responsibility for the management of the external borders' provision was included in the amended 2016 Regulation of both the Council and the Parliament 12 years later. This phrase has been added in line with Commission's proposal articulated with the 2015 Agenda on Migration that calls for a 'shared management of the European border'. Similarly, the 2004 Regulation did not make any reference to prevention or preventive measures. It is striking, though, that its 2016 revision lists the term prevention (or preventing/preventive) at least 13 times. Prevention refers to the proactive tackling of border control threats. So, it is more easily associated with securitisation, which constitutes one of Warsaw's border control assumptions, rather than Schengen's national security-oriented border control implemented by flexible practices. Moreover, comparing the 2006 Schengen Borders Code with its 2016 revision, it is possible to note a substantial rise in the use of the words 'risk' and 'security' across the 2016 document.[4] In a similar context, the concept of the pre-frontier area, which implements an expansion of the border area in neighbouring countries, in other words an extra-territorialisation, started to be officially used after

2008 (Commission, 2008), although now it is one of the most commonly used operational jargons (Regulation, 2019). Another indicative example is ETIAS, which reflects Warsaw's cultural elements, such as intelligence and information gathering. ETIAS was tabled according to a Commission proposal (2016b; Interviewee 3) signalising that Warsaw culture has become dominant not only among its practitiocratic border control community members. It is now embraced even by outsiders, such as the Commission.

The shift in attitudes manifests that the EU institutions have moved from Schengen to Warsaw. Although they did not plan or intend to replace Schengen, they contributed to its succession. In fact, the role of the EU institutions was rather pivotal. Without their input, Warsaw culture might never have existed. Indeed, the EU institutions created Frontex, namely the actor that developed and promoted this new culture. Next, despite any objections raised, especially during the agency's formative years (Colimberti, 2008: 37), the EU institutions allocated to Frontex additional powers strengthening its mandate, which led Frontex to consolidate its role in EU border control and in turn pursue more actively and with enhanced authority the new culture. Finally, the EU institutions allowed Warsaw's retention, namely the final step in the cultural evolution process. By agreeing and signing texts that included Warsaw's tenets, the EU institutions enabled Warsaw's institutionalisation and therefore assured its persistence over time. So, whilst Frontex was the actor that developed Warsaw culture, the EU institutions secured Warsaw's longévité.

Conclusion

What is Frontex's role in EU border control? To answer this question, this chapter provided an in-depth analysis of the impact of Frontex on EU border control. Rather than studying Frontex as an EU agency, remaining therefore at the surface of Frontex's function, it investigated its cultural impact, opting for a wider and more profound scrutiny of Frontex. The chapter argued that Frontex is impacting EU border control by promoting and pursuing a new cultural trajectory for border control, that is, Warsaw culture.

Drawing data from institutional discourse analysis, document analysis, interviews, empirical manifestations, and process-tracing evidence, the chapter attested the emergence of a new border control culture, labelled as 'Warsaw'. Unpacking this culture, it showed that Warsaw culture varies substantially from the other cultural trajectories, such as the previously dominant Schengen culture. Accordingly, it was shown that Warsaw culture is being pursued by a border control community composed of border control practitioners, namely Frontex and national border guards. It was also demonstrated that Warsaw culture is composed of certain shared assumptions and practices for border control. In particular, Warsaw perceives borders as national and EU external borders. Warsaw's border control assumptions include securitisation, technocracy, surveillance, intelligence, and extra-territorialisation/intra-territorialisation. Warsaw's border control practices are information gathering, multilateral cooperation, technology, and professionalisation.

Assessing Frontex's impact in the production of Warsaw's constitutive elements, the chapter revealed Frontex's causal role in Warsaw's development and consolidation. Actually, Frontex has produced Warsaw's border control community, its reference texts as well as border control assumptions and practices. This means that Frontex impacts on Warsaw culture by producing, promoting, and consolidating its components, which also constitute culture's essential conditions. Thus, Frontex is not just a decorative accessory. On the contrary, it has developed the Warsaw border control culture, leading therefore to a fundamental shift in EU border control. This shift refers to a different cultural approach pursued. Put differently, to the death of Schengen. But is this death irrevocable? Or has Warsaw culture started being challenged already?

Notes

1 The EU defines the integrated border management as 'national and international coordination and cooperation among all relevant authorities and agencies involved in border security and trade facilitation to establish effective, efficient, and coordinated border management at the external EU borders, in order to reach the objective of open, but well controlled and secure borders' (DG Home, 2022).
2 Finland is an exception to this as it developed its national Integrated Border Management Strategy in 2018.
3 See, for instance, the Nordic Co-Operation (2022) and Spanish-Moroccan relations in the 1990s (Lixi, 2017).
4 The word 'risk' is mentioned 25 times in the 2016 document, in comparison to 16 in the respective 2006 text and the word 'security' 51 times in relation to 35 in the 2006 document (Regulation, 2006, 2016b).

References

Adler, E. (2008) 'The Spread of Security Communities: Communities of Practice, Self-Restraint, and NATO's Post-Cold War Transformation'. *European Journal of International Relations*, 14(2), pp. 195–230.
Agnew, J. (1994) 'The Territorial Trap: The Geographical Assumptions of International Relations Theory'. *Review of International Political Economy*, 1(1), pp. 53–80.
Albahary, D. (2010) 'International Human Rights and Global Governance: The End of National Sovereignty and the Emergence of a Suzerain World Polity'. *Michigan State Journal of International Law*, 18(3), pp. 511–557.
Baldwin-Edwards, M. & Fakiolas, R. (1998) 'Greece: The Contours of a Fragmented Policy Response'. *South European Society and Politics*, 3(3), pp. 186–204.
Beach, D. & Pedersen, R.B. (2013) *Process-Tracing Methods: Foundations and Guidelines*. Ann Arbor: University of Michigan.
Bigo, D. (2000) 'When Two Become One: Internal and External Securitisations in Europe'. In: M. Kelstrup & M.C. Williams (eds.), *International Relations Theory*

and the Politics of European Integration: Power, Security and Community. London: Routledge, pp. 171–204.

Bigo, D. (2014) 'The (In)Securitization Practices of the Three Universes of EU Border Control: Military/Navy-Border Guards/Police-Database Analysts'. *Security Dialogue*, 45(3), pp. 209–225.

Buzan, B. (1991) *People, States and Fear: An Agenda for Security Analysis in the Post-Cold War Era*. Brighton: Weatsheaf.

Buzan, B., Wæver, O. & de Wilde, J. (1998) *Security: A New Framework for Analysis*. Boulder: Lynne Rienner.

Callovi, G. (1992) 'Regulation of Immigration in 1993: Pieces of the European Community Jig-Saw Puzzle'. *The International Migration Review*, 26(2), pp. 353–372.

Campesi, G. (2022) *Policing Mobility Regimes: Policing Mobility Regimes*. Abingdon: Routledge.

Carrera, S., Hernanz, N. & Parkin, J. (2013) 'Local and Regional Authorities and the EU's External Borders: A Multi-Level Governance Assessment of Schengen Governance and Smart Borders'. *Committee of the Regions*, Study, pp. 1–53.

Chillaud, M. (2012) 'Frontex as the Institutional Reification of the Link between Security, Migration and Border Management'. *Contemporary European Studies*, 2, pp. 45–61.

Colimberti, F. (2008) *Frontex: A Principal-Agent Perspective*. Munich: GRIN.

Commission of the European Communities (2002) 'Towards Integrated Management of the External Borders of the Member States of the European Union'. COM(2002)233.

Commission of the European Communities (2003) 'Presidency Conclusions: Thessaloniki European Council'. Available at https://ec.europa.eu/commission/presscorner/detail/en/DOC_03_3 (accessed January 2022).

Commission of the European Communities (2006) 'Reinforcing the Management of the Southern Maritime External Borders'. MEMO/06/454.

Commission of the European Communities (2008) 'Examining the Creation of a European Border Surveillance System (EUROSUR)'. COM(2008)68.

Council of the EU (2006a) 'Council Conclusions on Integrated Border Management'. 2768th Justice and Home Affairs Council Meeting.

Council of the EU (2006b) 'Frontex Feasibility Study on Mediterranean Coastal Patrols Network MEDSEA'. 12049/06.

Council of the EU (2018) 'European Council Conclusions: 28 June 2018'. Available at https://www.consilium.europa.eu/en/press/press-releases/2018/06/29/20180628-euco-conclusions-final/ (accessed April 2022).

Crawford, N.C. (2002) *Argument and Change in World Politics: Ethics, Decolonization and Humanitarian Intervention*. Cambridge: Cambridge University Press.

Csernatoni, R. (2018) 'Constructing the EU's High-Tech Borders: FRONTEX and Dual-Use Drones for Border Management'. *European Security*, 27(2), pp. 175–200.

DG Home (2022) 'European Integrated Border Management'. Available at https://home-affairs.ec.europa.eu/pages/glossary/european-integrated-border-management_en (accessed April 2022).

Dijstelbloem, H., van Reekum, R. & Schinkel, W. (2017) 'Surveillance at Sea: The Transactional Politics of Border Control in the Aegean'. *Security Dialogue*, 48(3), pp. 224–240.

EBCG Day (2014) 'ED4BG 2014 Integrated Border Management and its Way Forward'. Available at https://www.youtube.com/watch?v=MAydhOJlRns (accessed April 2022).

EBCG Day (2022) 'Home'. Available at https://ebcgday.eu/ (accessed February 2022).
Ekelund, H. (2014) 'The Establishment of FRONTEX: A New Institutionalist Approach'. *Journal of European Integration*, 36(2), pp. 99–116.
European Commission (2015a) 'A European Agenda on Migration'. COM(2015)240.
European Commission (2015b) 'A European Border and Coast Guard and Effective Management of Europe's External Borders'. COM(2015)673.
European Commission (2016a) 'Press Release: European Border and Coast Guard Agreed'. Available at https://ec.europa.eu/commission/presscorner/detail/en/IP_16_2292 (accessed January 2022).
European Commission (2016b) 'Stronger and Smarter Information Systems for Borders and Security Brussels'. COM(2016)205.
European Parliament (2021) 'Poland: MEPs Call for the Primacy of EU Law to be Upheld'. Available at https://www.europarl.europa.eu/news/en/press-room/20211014IPR14911/poland-meps-call-for-the-primacy-of-eu-law-to-be-upheld (accessed March 2022).
Frontex (2010a) *Beyond the Frontiers*. Warsaw: Frontex.
Frontex (2010b) *Frontex General Report 2010*. Warsaw: Frontex.
Frontex (2011) *Best Practice Guidelines on the Design, Deployment and Operation of Automated Border Crossing Systems*. Warsaw: Frontex.
Frontex (2012) *General Report 2012*. Warsaw: Frontex.
Frontex (2014) *12 Seconds to Decide*. Warsaw: Frontex.
Frontex (2015a) 'Fabrice Leggeri Takes the Helm at Frontex'. Available at https://frontex.europa.eu/media-centre/news/news-release/fabrice-leggeri-takes-the-helm-at-frontex-Z30Vu6 (accessed April 2022).
Frontex (2015b) *Frontex Annual Activity Report 2014*. Warsaw: Frontex.
Frontex (2015c) *Best Practice Operational Guidelines for Automated Border Control (ABC) Systems*. Warsaw: Research and Development Unit.
Frontex (2018) 'Stolen Vehicle Expert'. Available at https://frontex.europa.eu/media-centre/our-officers/stolen-vehicle-expert-c3dCwU (accessed March 2022).
Frontex (2019a) *Technical and Operational Strategy for European Integrated Border Management*. Warsaw: Frontex.
Frontex (2019b) 'Screening Expert'. Available at https://frontex.europa.eu/media-centre/our-officers/screening-expert-lV85qY (accessed April 2022).
Frontex (2021) *Annual Implementation Report 2020*. Warsaw: Frontex.
Frontex (2022) 'Information Management'. Available at https://frontex.europa.eu/we-know/situational-awareness-and-monitoring/information-management/ (accessed January 2022).
George, A.L. & Bennett, A. (2005) *Case Studies and Theory Development in the Social Sciences*. London: MIT Press.
Glouftsios, G. (2018) 'Governing Circulation through Technology Within EU Border Security Practice-Networks'. *Mobilities*, 13(2), pp. 185–199.
Græger, N. (2016) 'European Security as Practice: EU-NATO Communities of Practice in the Making?'. *European Security*, 25(4), pp. 478–501.
Guild, E., Carrera, S. & Geyer, F. (2008) 'The Commission's New Border Package Does It Take Us One Step Closer to a Cyber-Fortress Europe?'. *CEPS*, Policy Brief 154, pp. 1–5.
Haas, P.M. & Haas, E.B. (2002) 'Pragmatic Constructivism and the Study of International Institutions'. *Millennium: Journal of International Studies*, 31(3), pp. 573–601.
Horii, S. (2012) 'It Is about More than Just Training: The Effect of Frontex Border Guard Training'. *Refugee Survey Quarterly*, 31(4), pp. 158–177.

Huysmans, J. (2000) 'The European Union and the Securitization of Migration'. *Journal of Common Market Studies*, 38(5), pp. 751–777.

Jeandesboz, J. (2017) 'European Border Policing: EUROSUR, Knowledge, Calculation'. *Global Crime*, 18(3), pp. 256–285.

Jorry, H. (2007) 'Construction of a European Institutional Model for Managing Operational Cooperation at the EU's External Borders: Is the FRONTEX Agency a Decisive Step Forward?'. *CEPS*, Challenge Research Paper 6, pp. 1–32.

Jumbert, M.G. (2012) 'Controlling the Mediterranean Space through Surveillance: The Politics and Discourse of Surveillance as an All-Encompassing Solution to EU Maritime Border Management Issues'. *Espace, Populations, Sociétés*, 2012(3), pp. 35–48.

Kitchen, V.M. (2009) 'Argument and Identity Change in the Atlantic Security Community'. *Security Dialogue*, 40(1), pp. 95–114.

Léonard, S. (2010) 'EU Border Security and Migration into the European Union: Frontex and Securitisation through Practices'. *European Security*, 19(2), pp. 231–254.

Lixi, L. (2017) 'Beyond Transactional Deals: Building Lasting Migration Partnerships in the Mediterranean'. *Migration Policy Institute*, Report, pp. 1–22.

Longo, M. (2016) 'A "21st Century Border"? Cooperative Border Controls in the US and EU after 9/11'. *Journal of Borderlands Studies*, 31(2), pp. 187–202.

Lyon, D. (ed.) (2003) *Surveillance as Social Sorting: Privacy, Risk, and Digital Discrimination*. London: Routledge.

Marenin, O. (2010) 'Challenges for Integrated Border Management in the European Union'. *DCAF*, Occasional Paper 17, pp. 1–161.

Marin, L. (2011) 'Is Europe Turning into a "Technological Fortress"? Innovation and Technology for the Management of EU's External Borders: Reflections on FRONTEX and EUROSUR'. In: M.A. Heldeweg & E. Kica (eds.), *Regulating Technological Innovation: Legal and Economic Regulation of Technological Innovation*. Basingstoke: Palgrave Macmillan, pp. 131–151.

Nathan, L. (2006) 'Domestic Instability and Security Communities'. *European Journal of International Relations*, 12(2), pp. 275–299.

Nordic Co-Operation (2022) 'Organisation'. Available at https://www.norden.org/en (accessed April 2022).

Olsthoorn, P. & Schut, M. (2018) 'The Ethics of Border Guarding: A First Exploration and a Research Agenda for the Future'. *Ethics and Education*, 13(2), pp. 157–171.

Pallister-Wilkins, P. (2015) 'The Humanitarian Politics of European Border Policing: Frontex and Border Police in Evros'. *International Political Sociology*, 9(1), pp. 53–69.

Parkin, J. (2012) 'EU Home Affairs Agencies and the Construction of EU Internal Security'. *CEPS*, Paper in Liberty and Security in Europe 53, pp. 1–43.

Paul, R. (2017) 'Harmonisation by Risk Analysis? Frontex and the Risk-Based Governance of European Border Control'. *Journal of European Integration*, 39(6), pp. 689–706.

Philpott, D. (2001) *Revolutions in Sovereignty: How Ideas Shaped Modern International Relations*. Princeton: Princeton University Press.

Regulation (EC) No 2007/2004 of 26 October 2004 Establishing a European Agency for the Management of Operational Cooperation at the External Borders of the Member States of the European Union [2004, OJ L 349/1].

Regulation (EC) No 562/2006 of the European Parliament and of the Council of 15 March 2006 Establishing a Community Code on the Rules Governing the Movement of Persons across Borders (Schengen Borders Code) [2006, OJ L 105/1].

Regulation (EU) 2016/1624 of the European Parliament and of the Council of 14 September 2016 on the European Border and Coast Guard and Amending Regulation (EU) 2016/399 of the European Parliament and of the Council and Repealing Regulation (EC) No 863/2007 of the European Parliament and of the Council, Council Regulation (EC) No 2007/2004 and Council Decision 2005/267/EC [2016a, OJ L 251/1].

Regulation (EU) 2016/399 of the European Parliament and of the Council of 9 March 2016 on a Union Code on the Rules Governing the Movement of Persons across Borders (Schengen Borders Code) [2016b, OJ L 77/1].

Regulation (EU) 2017/2226 of the European Parliament and of the Council of 30 November 2017 Establishing an Entry/Exit System (EES) to Register Entry and Exit Data and Refusal of Entry Data of Third-Country Nationals Crossing the External Borders of the Member States and Determining the Conditions for Access to the EES for Law Enforcement Purposes, and Amending the Convention Implementing the Schengen Agreement and Regulations (EC) No 767/2008 and (EU) No 1077/2011 [2017, OJ L 327/20].

Regulation (EU) 2018/1240 of the European Parliament and of the Council of 12 September 2018 Establishing a European Travel Information and Authorisation System (ETIAS) and Amending Regulations (EU) No 1077/2011, (EU) No 515/2014, (EU) 2016/399, (EU) 2016/1624 and (EU) 2017/2226 [2018, OJ L 236/1].

Regulation (EU) 2019/1896 of the European Parliament and of the Council of 13 November 2019 on the European Border and Coast Guard and Repealing Regulations (EU) No 1052/2013 and (EU) 2016/1624 [2019, OJ L 295/1].

Regulation (EU) No 1052/2013 of the European Parliament and of the Council of 22 October 2013 Establishing the European Border Surveillance System (Eurosur) [2013, OJ L 295/11].

Saurugger, S. (2013) 'Constructivism and Public Policy Approaches in the EU: From Ideas to Power Games'. *Journal of European Public Policy*, 20(6), pp. 888–906.

Saurugger, S. (2014) *Theoretical Approaches to European Integration*. Basingstoke: Palgrave Macmillan.

Tholen, B. (2010) 'The Changing Border: Developments and Risks in Border Control Management of Western Countries'. *International Review of Administrative Sciences*, 76(2), pp. 259–278.

Trondal, J. & Jeppesen, L. (2008) 'Images of Agency Governance in the European Union'. *West European Politics*, 31(3), pp. 417–441.

Unisys (2014) 'Study on the Feasibility of the Creation of a European System of Border Guards to Control the External Borders of the Union: ESBG'. Final Report.

Vatta, A. (2017) 'The EU Migration Policy between Europeanization and Re-Nationalization'. In: S. Baldin & M. Zago (eds.), *Europe of Migrations: Policies, Legal Issues and Experiences*. Trieste: Università di Trieste, pp. 13–31.

Vaughan-Williams, N. (2008) 'Borderwork beyond Inside/Outside? Frontex, the Citizen-Detective and the War on Terror'. *Space and Polity*, 12(1), pp. 63–79.

Walters, W. (2002) 'Mapping Schengenland: Denaturalizing the Border'. *Environment & Planning D: Society & Space*, 20(5), pp. 561–580.

Wedeen, L. (2002) 'Conceptualizing Culture: Possibilities for Political Science'. *American Political Science Review*, 96(4), pp. 713–728.

Williams, M.C. (2003) 'Words, Images, Enemies: Securitization and International Politics'. *International Studies Quarterly*, 47(4), pp. 511–531.

Wilson, D. (2018) 'Constructing the Real-Time Border: Frontex, Risk and Dark Imagination'. *Justice, Power and Resistance*, 2(1), pp. 45–65.

Wolff, S. & Schout, A. (2013) 'Frontex as Agency: More of the Same?'. *Perspectives on European Politics and Society*, 14(3), pp. 305–324.

Wolff, S. (2012) *The Mediterranean Dimension of the European Union's Internal Security*. Basingstoke: Palgrave Macmillan.

Zaiotti, R. (2011) *Cultures of Border Control: Schengen & the Evolution of European Frontiers*. Chicago: University of Chicago Press.

Interviews

Interviewee 1: Frontex officer in Warsaw, 11 June 2018.
Interviewee 2: Frontex officer in Warsaw, 12 June 2018.
Interviewee 3: Frontex officer in Warsaw, 13 June 2018.
Interviewee 5: Border guard that had been seconded in Evros, 16 October 2018.
Interviewee 9: Police officer in Evros, 24 October 2018.
Interviewee 12: Border officer in Lampedusa, 11 May 2018.

7 Challenges to Warsaw culture

'[...] we are proud of our achievements.
At the same time, we are aware of the many challenges ahead'.
—Fabrice Leggeri, former Frontex Executive Director
(Frontex, 2022a)

Introduction

Border control may have been reinvigorated after Warsaw's cultural ascendancy, but this does not mean that it has ensured serenity. In fact, since Warsaw's institutionalisation as the dominant culture of border control in Europe, little has remained the same. In 2020, a new disease spread worldwide, infecting and killing millions of people as well as fundamentally changing the way we live and interact with other persons. The same year the EU lost one of its member states. Later, Europe's 'perpetual peace' shattered with the 2022 war in Ukraine. So, can Warsaw remain unscathed by these developments and continue savouring a dominant position in European border control?

This chapter reflects on Warsaw's present and future. In view of tectonic shifts at global, EU, and even national level that profoundly impact underlying understandings and policies about borders, the chapter assesses three challenges for the Warsaw culture and the border control policy community.

Initially, the chapter deliberates on why and how, just a few years after its institutionalisation, a culture can become unsettled. Then, it delves into Warsaw's challenges. The first challenge explored is Brexit and the English Channel's transformation from internal to external border. Although the UK was not part of the Schengen area and, therefore, not bound by Frontex's Regulations, Brexit still impacts the function and conception of EU borders by constructing a new external border between France and the UK. This means new symbolisms, new border regulations, and new border control actors managing border transactions. The next challenge discussed is the COVID-19 pandemic and its impact on borders' function. Can we still talk about a common border control culture, in the midst of divergent national

DOI: 10.4324/9781003230250-7

strategies for COVID-19 containment, including the closure of borders? The last section elaborates on 'hard' security, discussing the escalation of tensions and war threats between Greece and Turkey as well as the Russian invasion of Ukraine. In this war context, is there still room for border management considerations?

The chapter points out that these challenges may define both Warsaw's and Frontex's future course. Although they fall outside Frontex's remit, they challenge the governance of external borders and the dynamics of the border control policy community. So, if Frontex wants to retain its dominant position, it then needs to start addressing a wider border lexicon.

Unsettling a settled culture

Cultures are not fixed. They evolve, change (Swidler, 1986; Zaiotti, 2011; Lewens, 2015), or become replaced by other cultures depicting a cultural evolution. The first step for a cultural evolution, namely cultural variation, presupposes culture's contestation by alternative cultural paradigms.

According to Zaiotti, 'even when dominant a culture can still be contested' (2011: 24). After all, this is what happened with Schengen; after being formed, it challenged and, then, replaced Westphalia (Zaiotti, 2011). Later it was cast aside, ceding its dominant position in European border control to Warsaw culture, as described previously (Chapter 6). So, becoming dominant does not ensure a culture's longévité. Nor does it ensure a brièveté, namely a short life-span. Especially if the culture starts facing unexpected events or shocks. Then it can become seriously challenged and unsettled, initiating the process of cultural evolution that may lead to its eventual replacement by an alternative culture.

The emergence of alternative assumptions and practices, that is, an alternative culture of border control, which can earnestly challenge the dominant culture, may occur when the until then-dominant culture becomes unable to adapt to new circumstances or cannot explain and address effectively the problems preoccupying the policy community (Zaiotti, 2011: 32). Similarly, a fertile ground for the development of alternative cultures and the contestation of the dominant culture can be cultivated in unsettled times or, put differently, during an era of transformation or change (Swidler, 1986: 278–280; Martin, 1992), which paralyse the community's routines as well as question its underlying beliefs, evoking, at the same time, new meanings and actions (Scott, 2008: 39). In most cases, the challenges that arise to the dominant culture's authority are minimal. They can be easily contained without shaking taken for granted understandings and practices. In other words, without seriously unsettling its dominant position (Zaiotti, 2011: 24). Yet, a major challenge or a growing number of emerging challenges may shake a culture's hegemony, leading it, sooner or later, to fade into oblivion. So, a culture, even a dominant one, whilst seeking to remain relevant, is under constant pressure from both internal and external developments or stimuli (Berger, 1996: 326).

Hence, although the previous chapter attested that Warsaw has been 'coronated' the new dominant culture of border control in Europe after replacing Schengen, this does not mean that Warsaw's cultural life still dwells unchallenged. Both the world and Europe have changed since Warsaw's rise and institutionalisation in the 2010s. A number of significant events have occurred, challenging the notion and function of various institutions or concepts, such as the state, the EU, borders, security, and international cooperation. Would it have been possible for Warsaw culture to remain unscathed and retain its dominant position? Or is a new cultural evolution already underway, paving the way for Warsaw's replacement by another cultural trajectory?

Brexit and the English Channel

Following a referendum vote conducted on 23 June 2016, the UK officially left the then 28-member EU on 31 January 2020. The UK's withdrawal from the EU, commonly known as 'Brexit', terminated their 47 years of common course, triggering a new reality for both parts.[1] Undoubtedly, they did not enjoy a happy marriage, as their relations were characterised by a difficult and complex coexistence or by irreconcilable differences, with Brexit being the end result (Buller, 1995).

Actually, even before Brexit, the UK sought a path of differentiated integration, or better put, lax cooperation in certain policy fields as opposed to the deepening of EU integration. Adopting a 'pick-and-choose' approach (Mitsilegas, 2017), the UK obtained an opt-out regime in the Justice and Home Affairs policy area. Yet, the opt-out was not absolute. It opted in for certain instruments based on national interest grounds, such as the European Arrest Warrant and the Schengen Information System II (HM Government, 2013; Mitsilegas, 2017: 225–226). Notwithstanding the opt-in/opt-out scheme, the UK decided to terminate its EU membership.

A prominent slogan for the 'vote Leave' campaign in the Brexit referendum was 'to take back control of our borders' (Goodman, 2017). Regardless, the UK never abolished border checks at its internal borders with other EU member states. It never signed the Schengen Agreement. Neither did it become party to the 1990 Schengen Implementing Convention. It continued carrying out border controls at the UK-EU borders whilst being an EU member state. In fact, the Protocol of the Treaty on the Functioning of the European Union cites that the UK was 'entitled [...] to exercise at its frontiers with other member states such controls on persons seeking to enter the UK as it may consider necessary for the purpose (a) of verifying the right to enter the UK, (b) of determining whether or not to grant other persons permission to enter the UK' (TFEU, 2016: Article 1).

Whilst outside the Schengen area and thereby not bound by EU measures adopted, the UK showed from the very start a keen interest in Frontex. Within three months after the Commission's proposal for Frontex's establishment

(Commission, 2003), the UK announced its intention to take part in the agency, based on the Protocol listing that both the UK and Northern Ireland can 'at any time request to take part in some or all of the provisions of this Schengen acquis' (TFEU, 2016: Article 4). The UK's wish to participate in Frontex did not lessen even when both the Council and the European Court of Justice rejected the UK's demand.[2] Still, the UK succeeded in taking part in Frontex, albeit not fully.

Frontex's founding Regulation (2004) approved a framework of UK-Frontex cooperation by encouraging Frontex to 'facilitate operational cooperation' with the UK and providing Frontex's management board the liberty to decide about the UK's involvement in Frontex's operational actions on a case-by-case basis. A research on the management board's decisions, which are published in Frontex's website, manifests that the UK's numerous requests to participate in Frontex operations were always approved on the grounds that 'the UK has a pool of very experienced [...] experts [...] and their experience in cross border crimes presents an added benefit' for Frontex operations (Management Board, 2016). On that account, since 2006 UK experts were involved in Frontex joint operations, like Poseidon and Hera (Frontex, 2006: 11–12). Indicatively, in 2010 the UK participated in 5 Frontex joint operations at the air, land, and sea borders (Frontex, 2010). The UK even led joint return operations organised and financed by Frontex (Frontex, 2014). Moreover, UK representatives were invited to attend Frontex's management board meetings, whereas British staff were recruited at Frontex. In addition, the UK steadily contributed to Frontex's annual budget.

Ahead of the official Brexit date, the last Frontex Regulation enabled a framework of Frontex-UK cooperation by keeping the UK's ability to get involved in Frontex activities without, however, specifying the type of such activities (Regulation, 2019). In that context, in 2020, 31 UK experts were deployed as special advisors to Frontex operations. Considering their profiles, they included debriefing and screening experts as well as second-line officers (Frontex, 2021b: 23).

Currently, with the UK recently becoming a third country, the Frontex-UK relationship is still in process. Sticking to its preferred modus operandi for dealing with third countries, the management board has authorised Frontex's Executive Director to negotiate Working Arrangements with both the UK and Northern Ireland (Management Board, 2021). Although their finalisation is still pending, the UK has already participated in Frontex operational activities, such as operation '30 Days at Sea' conducted in 2021, manifesting therefore that, irrespective of Brexit, Frontex and the UK are not expected to become strangers. Most probably, neither will they become foes. After all, it was the UK that pushed for Frontex's involvement with external border checks in Gibraltar (Politico, 2021; UK Foreign Secretary, 2021), namely a UK overseas territory bordering Spain.

The UK's evolution into a third country transforms its borders with the EU member states from internal to external borders. Henceforth, they become a pole of attraction of Frontex's operational interest. In that respect, Frontex has already started monitoring EU-UK borders and particularly the English Channel.

The English Channel, the Channel in short, or 'la Manche' in French, constitutes a maritime border between the UK and France. Its border development reflects the Anglo-French antagonism or even rivalry between the two countries evolving from a trade route and a symbol of colonial power to a maritime frontier delineating national sovereignty (Morieux, 2016). It is a maritime area which constitutes part of the North Atlantic Ocean. It separates the southern coast of the UK from the northern coast of France. Its narrowest part is the Strait of Dover or in French 'Pas-de-Calais'. It marks the shortest distance, only 33.8 km, between the two countries, connecting the British port of Dover in the county of Kent with the French port Calais (House of Commons, 2019).

Crossing the border, passengers can use sea ferries or the undersea rail. Ferries to the UK depart every 30 minutes with a crossing time to Dover of around 90 minutes, whereas the crossing time via the undersea rail is 35 minutes. Either through ferries or rail, upon their arrival at the other side of the border, passengers are being subject to document and/or vehicle control conducted by border control officers.

The English Channel constitutes one of the busiest passages in Europe with very frequent connection lines as well as heavy passenger and vehicle traffic throughout the day and year. At the same time, it constitutes a sea passage not easy to navigate, due to high winds, frequent storms, strong water currents, and low water temperature (Morieux, 2016: 25). Despite its dangerousness, apart from being a very busy shipping lane, the English Channel has also become a leading route for irregular border crossers transiting through France and trying to enter the UK from continental Europe.

By the end of the 1990s, irregular border crossings through this route had increased significantly due to the crises in the Balkans and the war in Yugoslavia. A further rise in irregular migration flow was reported in the wake of the 2015 migration crisis, with approximately 6,000 people living close to the border in makeshift camps known as the 'Calais Jungle', seeking a chance to reach UK soil irregularly (The Guardian, 2015). During that period, most irregular border crossing attempts involved people, mainly from Eritrea and Syria, trying to enter the UK by ferry or by rail, hidden inside or under lorries and cars.

To prevent irregular border crossings and better monitor the border, the UK allocated additional funding for border control. It deployed more border control officers, detection dogs, and new border control tools, namely cameras, heartbeat, and carbon dioxide detectors, as well as enhanced the border's fencing with temporary security barriers and new razor wire fence constructions around the perimeter of the port (BBC, 2015, 2016).

These measures, coupled with tighter border control checks, led to a shift in irregular border crossings with more people attempting to traverse the

English Channel on small inflatable boats instead of ferries or the rail so as to limit the danger of detection before entering the British territorial waters and therefore avoid their return to France. Most irregular migrants are young adult males of Afghan, Syrian, Iraqi, and Iranian nationality, while UK data for 2022 show a surge in the number of Albanian nationals. The tariff for smugglers' services ranges from €3,000 up to €7,000 (VOA, 2021). According to the UK Home Office (2022), in 2021, 28,526 irregular border crossers were detected arriving on small boats, compared to 8,466 in 2020, 1,843 in 2019, and 299 in 2018. Proportionally, the number of irregular border crossing detections of migrants concealed in lorries or containers at UK ports dropped from 1,052 in 2018 to 665 in 2021 (UK Home Office, 2022).

Besides any measures implemented at national level, to address more efficiently irregular migration influx through this route, the UK turned to France. The two then-EU member states signed a series of bilateral agreements, including the 2003 'Le Touquet' Agreement that enabled the presence of border control officers in each other's sea ports. Later, they agreed to the establishment of a joint command and control centre in Calais along with the sharing of intelligence data (UK Home Office, 2015). After the Brexit referendum, the two countries signed the Sandhurst Treaty that included the allocation of additional funding from the UK to France. In parallel, a Coordination and Information Centre started operating, bringing together officers from the UK Border Force and the French 'Police aux Frontières' (UK Home Office, 2018). The following year, the UK and France signed a joint action plan (UK Home Office, 2019), which focused on 'combatting illegal migration involving small boats in the English Channel'. To that end, since 2020, both the UK Home Office and the UK National Crime Agency (NCA) have maintained a permanent presence in French territory, namely in Paris and Calais (UK Home Office, 2020a).

The English Channel has been on Frontex's radar since 2016. The agency's Risk Analysis Unit noted an increase in detections of irregular border crossers hiding in vehicles at the UK-France internal border (Frontex, 2016: 26–27). Approaching the date of the UK's departure from the EU, the number of secondary movements via this route became further elevated, leading Frontex to step up its situational awareness by starting to analyse both smugglers' modus operandi and border crossers' profile (2020a: 54).

After Brexit, the UK-France border was transformed from an internal EU border into an external border between an EU member state and a third country. Being an external border, it was officially added to Frontex's remit, as the agency manages 'the crossing of the external borders' and addresses any challenges encountered at those borders (Regulation, 2019). As a result, the English Channel does not represent just a permanent entry in Frontex reports. Instead, it now constitutes a Frontex operational area. Indeed, since December 2021, when its Executive Director characterised the English Channel as 'a matter of concern', Frontex has maintained an operational presence at the border. Frontex's presence includes the deployment

Challenges to Warsaw culture 153

of an airplane to foster aerial surveillance and support France in border patrolling (Frontex, 2021c). The aircraft is equipped with surveillance sensors and a thermal camera, whereas its crew can also be involved in SAR incidents (Frontex, 2022b).

The above illustrate that the English Channel is not a typical EU external border. On the contrary, it varies extensively from the borders analysed in the preceding pages, namely Evros and Lampedusa. A key difference relates to the direction of migration flows. For Evros and Lampedusa the attention is on irregular border crossers entering the EU from a third country; for the English Channel it is on those exiting the EU trying to enter a third country. Similarly, it varies from Warsaw culture, that is, the dominant culture of EU border control, as it does not share some of its constitutive elements. Accordingly, the English Channel's border management is not based on Warsaw's texts, such as the 2016 Frontex Regulation. On the contrary it is being guided by national and bilateral agreements between France and the UK.

Furthermore, by being nourished and functioning, until Brexit, as an internal border, the English Channel only partially embraces Warsaw's border assumptions. In particular, it shares the conception that borders are national territories. However, it has not conformed yet to the EU external border status. In the future, its official transformation from internal to external border may lead to its eventual EU external border construction. Yet, for this attribute to be elevated into an assumption requires time, which is needed so that internal border elements of its prior status can become progressively and effectively discarded.

Turning to Warsaw's border control assumptions, it is possible to identify elements of the Warsaw culture in the UK-France border. For instance, one can easily draw an emphasis on the gathering and production of intelligence manifested with the creation of information centres and the sharing of intelligence data. Also, there is a drive for surveillance evidenced with the deployment of radars and cameras for 24/7 border monitoring. Moreover, the blurring of the geographic location of the border constitutes a common feature at the UK-France border justifying the carrying out of border checks in the other country's territory.

A main distinction refers to securitisation. For Warsaw culture, irregular migration constitutes a security problem often encountered under an emergency requiring an urgent action. A 'secure the border' repertoire also exists in the English Channel's border management (Parker et al., 2021). Yet, here irregular border crossing is framed as a 'criminal activity' entailing 'criminal aspects' and requiring the dismantlement of 'criminal networks' through their 'criminal prosecution' (UK Home Office, 2019). Evidently, the focus is on 'illegal' and not 'irregular' migration. That attests a move towards criminalisation (Kostas, 2017) instead of securitisation.

Another variation between the English Channel and Warsaw culture is militarisation. Adopting a 'tough' stance (The Guardian, 2022), the UK decided to heavily involve the military in the English Channel. It could be argued that this involvement is linked to the English Channel's history and symbolism as a space of British heroic defence against Germany during World War II (Davies et al., 2021: 2315–2316). Another reason stems from the Royal Navy per se. The UK's Royal Navy is one of the most capable military forces worldwide. So, its mere presence at the English Channel's waters with large warships could have functioned as a migration deterrence. Until now, the Navy's participation in the English Channel's patrolling has involved the deployment of aircraft, drones, and maritime vessels (House of Commons, 2022: 10–11). Fostering militarisation, the UK appointed an ex-marine, Dan O'Mahoney, as a Clandestine Channel Threat Commander in order to coordinate all operational partners (UK Home Office, 2020b). Furthermore, implementing the UK's government plan, since April 2022, the Ministry of Defence has 'take[n] operational primacy for cross-channel counter-migration operations' (House of Commons, 2022: 12). So far, though, the Navy's takeover of operational command from the UK Border Force has proven futile, as the number of irregular arrivals has remarkably increased.[3] Irrespective of the operational appraisal of the Navy's deployment at the UK-France border, irregular or illegal migration does not resemble a military combat. Crossing the English Channel is not an act of war. As such, irregular border crossers or even smugglers are not soldiers. So, bringing in the military to combat irregular migration may lead to a change not only in the border control conduct but also in the border control's nature and meaning.

On border control practices, similar to Warsaw culture, border control at the UK-France border relies on information gathering and technology with the deployment of modern border control tools, such as drones and artificial intelligence (ComputerWeekly, 2022; House of Commons, 2022). A key distinction refers to cooperation. Border control in Warsaw culture is being implemented through multilateral cooperation. At the English Channel, though, border control is being managed bilaterally, that is, between the UK and France, mainly via bilateral agreements or direct communication and synergy between the two countries.[4] Frontex's involvement may alter this trend. Indeed, the first airplane deployed by Frontex at the English Channel was provided by Denmark (Frontex, 2021c). Still, both the UK and France maintain a strong national prism as a means to preserve their national sovereignty and fulfil their national ambitions for supremacy, thereby fuelling the Anglo-French antagonism. That was manifested, for instance, when France blocked a British company from providing border patrolling services in the English Channel on behalf of the EU (The Times, 2021) or when the French Prime Minister, Jean Castex, rejected a UK proposal for joint patrolling (Euronews, 2021).

Another difference between Warsaw and the English Channel relates to the border control community. In Warsaw, it is up to the border control practitioners, namely border guards, to pursue and promote Warsaw culture. That is not the case with the UK-France border. The border is being managed by various actors, such as police officers, national intelligence officers, military staff, and border guards. Deriving from different institutional sectors, they cannot perceive themselves as part of a community. Hence, it is harder to develop a common culture. At the same time, migration through the English Channel is an issue attracting public-wide concern and therefore continuous media coverage. That results in its politicisation. Besides being a main part of the Brexit campaign, the British Prime Minister and the French President often remark on the English Channel's irregular flows, even clashing publicly and trading accusations (The Guardian, 2021). So, the involvement of top-level officials or even the heads of the state reins in practitioners' input.

In short, the transformation of the English Channel from internal to external EU border after Brexit constitutes a fundamental change at the EU borders and, in turn, a challenge to Warsaw culture. By not being an EU external border, the English Channel was not part of the Warsaw culture. In other words, Warsaw was missing from the UK-France border. It still is, if we take into consideration that its external border status is a rather recent development. Yet, while it became a new part of the EU external borders, the border control actors of the English Channel brought their distinctive mentality regarding border management. This mentality challenges certain taken-for-granted logics for border control settled with Warsaw. Its peculiar nature and rather recent transformation from internal to external EU border do not allow the formulation of secure predictions about its future course and relationship with the Warsaw culture. Even so, one may wonder whether retaining its particular cultural elements in the future, the English Channel would represent a threat to the Warsaw culture. Is there any valid possibility of the English Channel functioning as an example for other EU borders and border control actors aspiring to copy any policies implemented there? Or is it more likely to be always considered a 'special case' being therefore dissociated from the EU border control system?

Borders amid the COVID-19 pandemic

On 31 December 2019, the World Health Organisation (WHO) became informed about cases of viral pneumonia in the Chinese city of Wuhan, for which the cause was unidentified. In just a month, similar cases were reported in 18 countries outside China, with scientists identifying a new coronavirus and naming it 'COVID-19'. A surge in cases prompted WHO's Director-General to declare, first, a public health emergency, and, then, the outbreak of a pandemic (WHO, 2022a). Soon Europe became the virus' epicentre, reporting a high number of cases and mortalities.[5]

'To slow down the spread of coronavirus and protect the health and well-being of all Europeans' (European Commission, 2022), member states closed their borders, imposing mobility restrictions including travel bans to and from risk areas. Soon, total entry bans came into effect in an attempt to combat further increase in the number of infections. In this context, after the first cases reported in Wuhan, in the beginning of 2020, the Chinese authorities confined 11 million people to their homes for 76 days. Successively, countries started to seal their borders to travellers arriving from China as a means to keep away the virus. North Korea was the first country to adopt such a policy, in January 2020, with Afghanistan, USA, Singapore, Russia, and Australia following its lead the next month. Amid the rise in COVID-19 cases detected outside China, mobility bans and border closures became amplified, with countries adopting a stricter policy for cross-border mobility as a means to contain the virus' spread. Travel restrictions imposed included border closures, foreign nationals' entry or exit bans, visa restrictions, flight suspensions from selected geographic areas according to their epidemiological situation, self-isolation, or even quarantine for 14 days as well as new entry requirements, such as mask wearing, negative COVID-19 test, and later vaccination proof.

In Europe, France was, on 24 January 2020, the first country to report a COVID-19 case in its territory and the first death on 15 February. Next, cases were announced by Germany, Finland, Italy, Sweden, Spain, and the UK. Parallel to the detection of COVID-19 cases in various EU countries, the road of travel bans became a foregone conclusion. Austria became the first country to reintroduce border checks at its internal borders with Czechia and Slovakia, starting from 3 February 2020, whereas Sweden imposed a new entry requirement of presentation to border control authorities of a negative test result so as to enter the country. After that, many countries started to impose mandatory self-isolation upon arrival, such as Estonia and Lithuania.

Travel restrictions do not solely refer to an administrative change in entry requirements or a shift in the protocol for entering the country, asking for additional travel documents. Most importantly, COVID-19 led to a change in the nature of borders. Suddenly, intended entrants started to be categorised and handled according to epidemiological data. Each entry was not any more treated as a cross-border issue or even migration concern. Instead, it was handled as a possible threat to public health, given that the virus was being transmitted from person to person at a rapid rhythm. In this context, even nationals entering their country of origin or residence became subject to travel restrictions. All entrants, irrespective of their status, origin, travel documents, or reason for travel were treated as potentially infected with a contagious disease and therefore a danger to national security.

In this environment, even internal EU borders became affected, with member states reintroducing border controls. The Schengen Borders Code allows member states to temporarily reintroduce border control at the internal borders in the event of a 'serious threat to public policy or internal security' (Regulation, 2006, 2016). Drawing on that, on several occasions, various member states have chosen to reintroduce border controls for a limited period of time, such as Germany in 2015 during a G7 Summit, Poland during the Euro 2012 Football Championship, as well as Denmark, Austria, Sweden, and France in different time-spans due to terrorist threat. Before COVID-19, there was only one precedent of extended periods of border controls at the internal borders of multiple countries. That was the 2015 migration crisis. The start was made by Germany, which, in September 2015, announced the reintroduction of checks at its land borders because of a large-scale influx of persons seeking international protection. This marked a chain reaction with different countries adopting the same decision, such as Austria, Slovenia, Hungary, Sweden, Norway, Denmark, and Belgium.

So, the temporary reintroduction of border control at internal borders is not rare. Rather, it constitutes a strategy implemented by different EU member states on divergent occasions. With COVID-19, though, this strategy became amplified through more than half of the EU member states reintroducing border controls at their internal borders for extended periods of time. In fact, 14 EU member states, joined by Schengen's Norway, Iceland, and Switzerland, reintroduced border controls at their internal borders due to the COVID-19 health emergency. That decision invoked a shift both in the nature and status of internal borders. By establishing border controls, they were equated with external borders.

Although the Schengen Borders Code does not list public health as a ground for the reintroduction of temporary border controls, still the spread of the virus coupled with the alarmingly high hospitalisation and Intensive Care Unit admission rates, and later deaths, led both the member states and the European Commission to handle COVID-19 as 'an extremely critical situation' during which member states could 'reintroduce border controls as a reaction to the risk posed by a contagious disease' for the protection of public health (European Commission, 2020a).

At the same time, on external borders, the European Commission advised member states to perform health checks on all persons entering the Schengen area, even EU nationals, as well as to prohibit persons infected with COVID-19 or exposed to the virus from travelling (European Commission, 2020a). On the next day, 17 March 2020, EU leaders agreed on an EU entry ban, namely to restrict non-essential travel to the EU, signalising a total and generalised Schengen closure (van Eijken & Rijpma, 2021: 36). A few months later, the Council recommended the adoption of common criteria in relation to the restriction of mobility on public health grounds. Emphasis was placed on the European Centre for Disease Prevention and Control's mapping of risk areas, which referred to a categorisation from 'green' to 'red' according to case notification rate. On that account, valid travel documents were not any more enough to enable entry into or even travel to an EU country.

So, COVID-19 led to a hardening of borders, both internals and externals, implemented via mobility restrictions or even refusals of entry. Gradually, new tools were deployed at the border to assist health monitoring, like thermo-vision or contactless thermometers as well as thermal cameras for immediate body temperature screening on arrival in French and Portuguese airports. In this context, the protection of public health through the prevention of COVID-19 spread, by refusing entry or isolating persons potentially infected with COVID-19, became added to border guards' duties and, in turn, to their role at the borders. In parallel, the COVID-19 outbreak manifested diverging measures adopted at national level. Besides the reintroduction of border control at different internal borders, there was inconsistency in risk assessments as well as different entry requirements. For instance, Greece permitted entry to the country by persons vaccinated even with vaccines not approved by the European Medicines Agency, such as the Russian Sputnik or the Chinese Sinovac and Sinopharm, whereas Germany did not. Similarly, in March 2022 Greece was alone in enacting a new mandatory protocol, requiring all travellers, even Greek nationals, to fill in, before their departure to Greece, a Passenger Locator Form entailing personal data and travel information. Hence, despite any coordination attempts at the EU level, during the COVID-19 crisis, EU member states accentuated their national policies by adopting divergent and even contradictory strategies to tackle the spread of the virus.

Regarding irregular border crossers, the COVID-19 outbreak became accompanied by a significant decrease in the number of arrivals. Before the pandemic, in 2019, 123,664 irregular border crossers arrived in the EU Mediterranean countries. That number dropped to 95,774 the next year (UNHCR, 2022a). Actually, during COVID-19, countries closed their borders not only to travellers but to all arrivals. For instance, both Malta and Italy declared their seaports unsafe, not allowing the landing of migrant rescue boats until the end of the health emergency. Implementing a mobility restriction policy, asylum seekers and irregular migrants, upon their entry, were being detained and placed in quarantine, suspending therefore their access to the asylum procedure (Greek Council for Refugees, 2021). In Italy, most newcomers were spending a mandatory quarantine period on board ships in the open sea before being transferred to reception centres. In Greece, besides quarantine to newcomers, the lockdown in structures hosting refugees and migrants continued to be in place, although the country had its borders open for tourists. These restrictions evidently had an impact on flows. Prospective irregular border crossers, to avoid being stranded in quarantine sites for an indefinite period of time, without access to the asylum procedure and NGO support, most likely opted to defer their attempt to cross the borders until mobility restrictions were eased.

Aside from — or maybe due to — its impact upon border flows and global mobility, COVID-19 also affected Frontex's work. Deployed officers at Frontex joint missions as well as staff at Frontex's headquarters became infected with the virus. As a result, they had to complete a mandatory self-isolation period. Having a number of inactive staff made a dent in Frontex's operational activities (Frontex, 2021d: 34). On top, new recruitment schemes became delayed, as the conduct of medical tests was not possible for more than three months (Frontex, 2021d: 35). Furthermore, various Frontex activities were held virtually, such as unit meetings, interviews for new staff, and conferences, like the International Conference on Biometrics for Borders. In parallel, safety and sanitary rules became established, involving arrangements for teleworking accompanied by a reduced presence of staff physically present in the office (Frontex, 2021d: 34) as well as temperature checks for deployed officers before every shift. Apart from the conduct of work, the virus also affected Frontex's content of work. Border guards soon came across a rise in fraudulent supporting documents, such as proof of vaccination or negative testing as well as smuggling of medical products and protective equipment (Frontex, 2021a: 7).

In the COVID-19 era, Frontex tried to reassert its key position in EU border control by clarifying that it 'plays a key role in ensuring effective protection of the external borders of the EU not only against cross-border crime but also against health threats' (Frontex, 2020b). Following that, Frontex refined its mandate, expanding the list of threats that it tackles and, in turn, its scope of action. Characterising it 'an unprecedented challenge' with unknown consequences, the agency called for a reorientation of border management towards health screening (Frontex, 2020a: 57). To do so, it pursued to regularly monitor and inform about restrictions to cross-border mobility. Also, it commenced assessing the impact of the virus on border and irregular migration flows. Building on its new scope of action, it co-organised a regional workshop on health and safety in border management as well as established a COVID crisis cell to constantly inform about the development of the pandemic situation.

Hence, the COVID-19 pandemic shaped a new reality both at the borders and for the border control actors, such as Frontex, redirecting their focus and work methods. Inserting new routines, such as regular health checks, self-isolation, and face masks, border guards came across and started to develop new behaviours and working habits. Yet, the question is whether this new environment triggered by COVID-19 clashes significantly with Warsaw's assumptions and practices.

A significant change in EU borders relates to the reintroduction of border checks at internal borders. Putting forward the national aspect of borders, borders' internal/external distinction became irrelevant. So, Warsaw's twofold consideration of borders as national and EU external borders was substituted by national borders' prominence coupled with their simultaneous hardening.

On border control assumptions and practices, certain new trends initiated amid the pandemic seem to complete Warsaw's tenets rather than clashing with them. A prominent example is securitisation, taking into account that COVID-19 was framed in most cases as a health emergency posing an existential threat to both global and national security, justifying, as a result, the suspension of rights and pursuit of extraordinary and exceptional measures, like border closures, mandatory testing, and quarantines, to combat the virus and protect the public health (Kirk & McDonald, 2021). Moreover, to facilitate movement across borders, an EU Digital COVID Certificate was established, which became a travel requirement and therefore an essential document to be presented upon arrival to border guards. To verify its authenticity, border guards scan it using digital devices. The inclusion of new digital forms and devices clearly coincides with Warsaw's assumption of technocracy. Furthermore, the need for physical distancing reinforced the use of technological tools, such as thermo-vision thermometers and thermal cameras for mass temperature screening. In that context, there are even suggestions to turn to artificial intelligence as a way to minimise physical contacts (Eržen et al., 2020). Similarly, the collection of data, including sensitive health information, attests a focus on information gathering. In the same spirit, the method of contact tracing manifests the diffusion of both surveillance and intelligence whilst seeking to monitor not only the entrant but also his/her closed contacts. Now, multilateral cooperation represents a conflicting case. At the outset of the pandemic in Europe, member states pursued national strategies, having in mind solely their national interests. On that account, borders closed, different measures started to be implemented, for example, regarding the valid entry requirements and quarantine period after entry, and a war for masks or other medical equipment was waged, with many EU member states opting to negotiate bilaterally with third countries in order to avoid supplies shortages. Yet, showing their solidarity, when Italy became heavily hit by the virus, European countries transferred medical staff and equipment to northern Italy. Furthermore, proving multilateral cooperation, several EU countries relocated a number of COVID-19 patients to neighbouring countries' hospitals, such as in Italy and Germany.

The border control community, despite being challenged by new actors at the borders, seems to be retaining its key position and border control practitioners' composition. To be sure, border control started to be conducted based on entry criteria set by the epidemiological situation. On that account, member states were consulting the European Centre for Disease Prevention and Control's analysis of risk areas, whereas medical staff, such as doctors and nurses, became a permanent presence at the borders so as to perform COVID-19 tests on entrants. Their involvement with border control, though, is indirect. The European Centre for Disease Prevention and Control or any medical staff present at the border are not the ones performing border control. That continues to be border guards' prerogative and duty.

Besides the emergence of new actors, the COVID-19 predicament also sparked a return to the national-level decision-making, especially during the first months of the pandemic signalising the concomitant disruption of EU level coordination. The return to the nation-state logic during an emergency situation could have wider and lasting implications threatening in turn the viability and future of the Warsaw culture. However, Warsaw's tumble does not seem prone to occur, given that the border control practitioners, which compose Warsaw's border control community, remained actively involved during the health emergency acquiring new duties and functions in the sake of the protection of insiders' health. In fact, the border control conduct became adjusted returning now to a border control normality.

In sum, COVID-19, on the one hand, completed and on the other, challenged Warsaw's tenets for border control. The use of technology, the collection of information, intelligence, and the protection of the 'insiders' from the 'outsiders' are also evident in Warsaw's culture of border control. But, the management of this health emergency has also brought a catholic sealing of the borders, new travel requirements, national policies, mobility based on epidemiological criteria, and the use of devices to check entrants' health. All these form a new border reality. Irrefutably, the emphasis on sanitary controls and biopolitical logics (Foucault, 2008) did not occupy a prominent place in the EU border lexicon before COVID-19, despite the world having witnessed various pandemics, such as Ebola, SARS, yellow fever, and cholera. Actually, two decades ago, Saudi Arabia was the first and only country to impose an entry ban on people arriving from SARS-stricken regions in South East Asia. Moreover, vaccination against yellow fever is referred to as a recommended or in some cases mandatory precondition for travel to some countries in South America or Africa, such as Uganda. The sporadicity of these cases cannot be compared to the global fury triggered by COVID-19. To slow down the virus' spread, the majority of EU countries sealed their borders and/or imposed mandatory quarantine for all entrants. After gaining access to vaccines, an EU COVID Certificate was created and became a mandatory travel requirement to EU countries. It remains open whether these new rules initiated amid COVID-19 along with the focus on health safety, as new infectious diseases are diagnosed, like the spread of monkeypox since May 2022, will remain in the post-pandemic era, or whether border guards will return to the status quo ante, performing their duties in a business-as-usual mode.

Borders and hard security

In the 2000s and 2010s, the European continent was savouring a perpetual peace with war being a non-imaginable scenario after recovering from the wounds of the wars in the Balkan region during the 1990s. Sadly, the start of the 2020s retriggered hard security considerations, depicting emphatically that war has not become obsolete in Europe. War scenarios resurged in the midst of the heightened Greek-Turkish juxtaposition, while, later, Russia's invasion of Ukraine brought a war to the EU's doorstep. These hard security developments put into question not only Warsaw culture, but the wholesale conception of borders as spaces for cross-border management of flows, as opposed to military frontiers.

Greek-Turkish row: The 2020 Evros border crisis

The historically problematic Greek-Turkish relations, filled with continuous disputes and numerous — yet failed — attempts for rapprochement, became even more complicated during 2020. Both sides of the Aegean, instead of deescalating tension and returning to their normal bad neighbourly coexistence, chose to further fuel it both in words and deeds, leading to the 2020 Evros border crisis.

Following months of increasingly strident rhetoric, including taking different sides in Libya's civil war, there ensued a series of actions perceived as deliberate provocation by both parties. The most potent provocation for Greece was Turkey's surprise delineation of an Exclusive Economic Zone with Libya in November 2019 and, in turn, Turkey's conduct of seismic surveys in contested waters, while for Turkey it was the signature of the Eastern Mediterranean (EastMed) natural gas pipeline agreement by Greece, Israel, and Cyprus in January 2020.

The crisis further escalated, with even war bells ringing rather loudly, when, amid the outbreak of the COVID-19 pandemic in Europe and after the death of 34 Turkish soldiers in the Syrian region of Idlib, the Turkish President announced, on 27 February 2020, that Turkey, hosting at the time more than 3.5 million Syrian refugees (UNHCR, 2022b), would cease to conduct border control on persons exiting the country and entering Europe, thereby opening its borders. His announcement sparked an unprecedented influx of persons gathering at the Greek-Turkish land border in Evros and, as a result, a stand-off at the border between Greece and Turkey. In fact, in just a few hours, the news of Turkey's border opening spread rapidly via WhatsApp, leading thousands of people to the neutral border zone between Greece and Turkey, mostly to Kastanies BCP, in an attempt to enter the EU. Despite the initial surprise of the Greek side over the sudden accumulation of people at the border, Greek border officials stationed there, standing shoulder-to-shoulder with police and military officers as well as civilians, tried to repel the large-scale influx, fighting 'man-to-man' (Ekathimerini, 2021). Using water cannons, tear gas, and sending automated texts warning 'not try to illegally cross the Greek border' in different languages, Greek authorities stated that they would protect the border at any cost, at the same time, blaming Turkey for instrumentalising migration and breaching the 2016 EU-Turkey Statement (Greek PM, 2020). Soon, the hashtag #Greece_under_attack started trending fuelling a war on social media between supporters of Greece and Turkey.

To deter people from attempting to cross the border, Greece declared that all irregular entrants would be deported to their countries of origin as well as suspended the processing of asylum applications. Moreover, it began announcing on a daily basis the number of averted irregular entries to the Greek soil. For its part, Turkey was refuting their blocking and accused Greece of excessive violence, even Nazi tactics, against migrants

(EURACTIV, 2020). Amid shooting incidents and the transfer of military officers, even special forces, at both sides of the border, fears of an escalation from a clash with migrants into an armed conflict between the two countries did not take long to be uttered. The war-like scenery demarcating the so-called 'battle of Evros' continued even after the visit of the 4 Presidents of EU institutions at the Greek-Turkish land border[6] as well as a phone call from the Greek Prime Minister to the US President, Donald Trump, briefing him on the Evros situation.

In the wake of mass movement towards the Greek-Turkish land border, Frontex became 'on high alert' (Politico, 2020). In the early days of the Evros border crisis, the agency, besides closely monitoring the border, transferred additional equipment and border officers to Greece. Later, it became more actively involved on the ground. Responding to the Greek request for immediate assistance, Frontex activated its rapid border intervention mechanism approving the conduct of RABIT operations at Greece's land and sea border with Turkey. The 2020 Evros RABIT operation started on 11 March and lasted, after repeated extensions, until 31 October 2020.

The Evros border crisis started to be deescalated one month after its outbreak, namely in 27 March 2020, when Turkish authorities commenced the evacuation of migrants from the border area. Yet, Greek-Turkish relations remained strained. In the following months, fuelling the tension between the two countries, Greece signed Exclusive Economic Zone agreements with both Italy and Egypt as well as announced the expansion of its territorial waters in the Ionian Sea from 6 to 12 miles, admonishing about its intent to exercise a similar right also in the Aegean Sea, despite this qualifying as a 'casus belli', namely a cause for war, for Turkey.[7] The increased presence of naval forces of both countries in the Aegean Sea even led to the collision of the Turkish frigate 'Kemal Reis' with the Greek 'Limnos' in August 2020, whereas France, worrying about the non-decreasing tension, boosted its military presence in the region.

The above convey a long period of escalating friction between the two neighbouring countries, stirring their border control conduct, with Evros becoming the 'topos' of a new border crisis. During that period, Warsaw culture became side-lined, as there was no room for border management. Conversely, all efforts were directed towards border protection. Border management and border protection differ substantially; they conceive and treat borders differently. Actually, amid the Evros border crisis, the Greek-Turkish border was mostly perceived as a national frontier and a European shield (Greek PM, 2020), rather than a border crossing point. In fact, all legal cross-border mobility halted, stripping then the border of one of its key functions and border management characteristics.

Moreover, during that emergency, the border became militarised. That can be extracted by the transfer of military officers to Evros as well as the use of military tactics against the migratory influx, like the conduct of military drills with live ammunition or the deployment of military equipment.

Actually, new military equipment, such as armoured security vehicles donated by the US Army to Greece, has become a permanent fixture at the Greek-Turkish borders, adding, in turn, to their militarisation. Furthermore, in the official discourse of both countries, there were clear threats of using armed force against each other to protect their national interests and sovereignty. For instance, during his visit to Evros, the Greek Prime Minister (2020) proclaimed that '[his] duty is to protect the sovereignty of [his] country', whereas the Turkish President declared that he would not allow Greece 'to obtain unjust gains for itself' (The Guardian, 2020). Although border militarisation is not a new trend but an ever-present aspect at the Greek-Turkish border, the 2020 Evros border crisis displayed militarisation differently. As discussed in the analysis of Evros border (Chapter 5), its militarisation emanates from the border assumption of Evros being a national territory. In this context, the 'threat', and therefore the only reason to be militarised, is Turkey. After all, the reason why Greece had chosen to lay landmines at Evros border was to protect its territorial integrity against a Turkish invasion. Yet, during the 2020 events, the overt Evros battle was not between the two countries, namely Greece and Turkey. Instead, the warring parties were, on the one side, the Greek forces preventing entry to the country; on the other side, irregular migrants trying to enter the Greek soil. Interestingly, migrants were not threatening Greece's territorial integrity.

In reality, migrants became instrumentalised by both countries. First, by Turkey, which 'actively encouraged migrants and refugees to take the land route to Europe through Greece' (European Commission, 2020b: 7). Its aim was to strike a new EU-Turkey deal through having more leverage, as well as to divert public attention from the losses in Syria to Evros. Then, by Greece, which, on account of the Evros battle, closed its borders, suspended entrants' access to the asylum procedure, and started implementing Evros' fortification by enhancing its border fence as well as deploying more border guards and new border control tools, like sound cannons. In this context, incidents of alleged push-back practices in Evros increased (CoE, 2020).

Hence, the 2020 Evros border crisis points towards the development of a new reality at the border. Indeed, the war context shaped during the Evros battle brought a substantial militarisation accompanied by new border control tools and tactics. During that period, irregular border crossers were trying overtly to enter Greece in large numbers without using smugglers' services. In that context, intelligence, which constitutes one of Warsaw's border control assumptions, lost its value for the border control conduct. Moreover, at odds with Warsaw's spirit of extra-territorialisation/intra-territorialisation, amid the Evros crisis, the border-line became again relevant with Greek border officials being lined up at the border to prevent all entries to the Greek territory. Regarding border assumptions, the national context gained momentum with the border being treated as a battlefield and a shield from an orchestrated attack against Greece's national sovereignty. In fact, border protection became the chief preoccupation for border control actors.

Although some time has passed since the escalation of the crisis, some of these elements continue to shape Evros border control conduct, raising questions about Warsaw culture and whether it has become irrelevant to Evros border under the weight of hard security considerations. The verdict remains open. Irrefutably, though, Evros was not a sui generis case. The next year, Belarus, pursuing an analogous stance with Turkey's role in Evros, attempted to instrumentalise or, as others put it, weaponise migration as a political retaliation for EU sanctions (Foreign Affairs, 2021). Trying to respond to the mass migration flow, the EU-Belarusian border became militarised, as neighbouring to Belarus EU states, namely Poland, Latvia, and Lithuania, deployed troops and conducted military exercises near the border to function as deterrence.

Despite the novel situation encountered during the 2020 Evros crisis or even the 2021 EU-Belarusian stand-off, Warsaw culture was not totally absent from the border. Surveillance, technology, multilateral cooperation with the deployment of border officers from other EU countries are some of Warsaw's characteristics heightened during both border crises. Asserting that, Frontex coordinated RABIT operations during both crises, deploying border officers and technical tools to assist with the protection of the EU external borders. So, besides the war context, Warsaw's spirit continued to reside at the border, shaping as a consequence the pursuit of border control.

War in Europe: The 2022 Russian invasion of Ukraine

The Greek-Turkish stand-off in Evros triggered fears of escalation with direct military confrontation between the two countries leading to the outbreak of war in Europe. That fear became a bitter reality two years later when Russia invaded Ukraine, bringing war onto the EU's doorstep. Although Ukraine was part of the Soviet Union, since the latter's dissolution, the Russo-Ukrainian relationship had been floating from turbulent to aggressive. Kiev's aspiration to head towards a pro-western path, including its goal for NATO and EU membership, despite Moscow's strong opposition, led Russia and Ukraine to start drifting apart. Their relations worsened, when, in the wake of the 2014 Maidan protests, the Russian-backed President, Viktor Yanukovych, fled the country.[8] A few months later, in March 2014, Russia annexed the Ukrainian peninsula of Crimea in accordance with the result of a disputed referendum (UN, 2014). In the aftermath of Crimea's annexation, Russo-Ukrainian relations sank to an all-time low. That changed 8 years later. Russia's military invasion of Ukraine not only caused their bilateral relations to hit a new low; most importantly, it rendered them mortal enemies in warfare.

After several weeks of assembling troops on the Russo-Ukrainian border and having officially recognised the independence of Moscow-backed rebel regions Donetsk and Luhansk, on 24 February 2022, Russia launched the conduct of a 'special military operation' against Ukraine. This was a full-scale military attack on Ukraine's north and east involving land, air, and naval forces, with Russia's nuclear arsenal placed on a higher state of alert. Despite mediation attempts by third countries and international organisations as well as the imposition of severe sanctions against Russia, the war continued to wage for several months.[9] Various Ukrainian cities were bombed, even Kiev, whereas the Russian troops met fierce resistance from the Ukrainians. The war led to more than 6,000 civilian casualties, including 3,000 deaths (OHCHR, 2022), and the displacement of more than 5 million people (UNHCR, 2022c). Most were women and children that were fleeing to EU countries bordering Ukraine, namely Poland, Hungary, Slovakia, and Romania.

That was one of the fastest and largest movements of people towards the EU territory, with the EU having to handle a new refugee crisis. That crisis, though, does not bear any resemblance to the crisis of 2015. Under a wave of solidarity towards the Ukrainian people, Ukraine's neighbouring EU countries opened their borders and ceased border controls.[10] Even COVID-19 proof of vaccination stopped being an entry requirement for those fleeing the Russo-Ukrainian war and entering the EU soil. Yet, the now frontline EU countries did not have prior experience in dealing with a mass flow of asylum seekers. The 2015 migration crisis targeted mainly Greece and Italy, whereas Hungary and Poland, adopting an anti-migrant stance, refused to take in applicants for international protection under the EU relocation scheme. So, the reception facilities in these countries were rather limited. Showing solidarity, the EU member states unanimously agreed to share the burden of reception by activating, for the first time since its 2001 adoption, the Temporary Protection Directive, granting immediate temporary protection to those fleeing the war in Ukraine in any EU country (Council Directive, 2001), namely entitlement to stay, move, and work across the EU without first being granted refugee or subsidiary protection status.

The unprecedented cross-border flow towards the Eastern Borders route due to the war in Ukraine also mobilised Frontex. The very next day of the Russian invasion, Frontex announced that it was monitoring on a 24/7 basis the situation in Ukraine and expressed its readiness to assist EU member states even with crisis response teams already being activated (Frontex, 2022c). On the ground, Frontex deployed additional border officers and equipment to EU member states bordering Ukraine. Furthermore, it warned about a possible increase in smuggling of weapons and people. Also, it coordinated a joint operation in a third country, namely in Moldova, and organised its first humanitarian return flight assisting non-Ukrainian citizens that had fled the war to reach their country of origin. In synergy with other EU agencies, namely Europol and Eurojust, Frontex also contributed to EU investigations on sanctions imposed in relation to the Russian invasion. In particular, it was tasked to scrutinise whether the persons crossing EU's external borders fell under the scope of the imposed EU sanctions.

In the war context described, the status and function of borders altered significantly. The EU-Ukrainian borders stopped being an us/them or inside/out demarcation. Border controls were lifted, and therefore reaching the border connoted entering the EU territory and being welcomed to the EU member states. Most importantly, leaving Ukraine practically equated with escaping war and surviving. Indeed, Warsaw's binary assumption of borders as national and EU external borders became irrelevant. Likewise, upon the abolition of border checks, most of Warsaw's border control assumptions and practices, such as intelligence and surveillance, were cast aside. At the same time, others were heightened. For instance, multilateral cooperation and information gathering prevailed against national seclusion, with EU member states agreeing to share the burden of reception. In that context, Frontex's joint operation being conducted at the EU-Ukrainian border, namely Terra 2022, became enhanced with additional border officers sent to assist with the mass inflow of persons and their swift registration. In total, 520 Frontex officers were deployed at the EU's eastern borders (Frontex, 2022d). That means that border guards continued being present at the border. But rather than preventing border crossings, they were actually facilitating them. In this regard, border guards' duty converted from checking border crossers' travel documents to providing assistance, including porter tasks, to people, mostly women and children, fleeing the conflict zone.

Irrefutably, in the midst of war, the meaning of borders and their function alter. In fact, even the threat of war can impact the border control conduct, as attested in the 2022 Evros crisis. So, hard security considerations can challenge Warsaw's spirit of border control. In the two cases analysed, though, namely Evros and Ukraine, hard security did not awaken similar characteristics. During the 2022 Evros crisis, the border became an impenetrable fortress; during the 2022 war in Ukraine, all entry requirements were lifted with borders becoming open to all entrants. This variation renders the prospect of Warsaw's contestation far more difficult. After all, even amid hard security considerations, Warsaw was not absent from the borders. Multilateral cooperation, which constitutes a Warsaw practice of border control, was a common trait encountered in both cases. The same applies to the border control practitioners, namely border guards and Frontex staff, which compose Warsaw's border control community. They remained at the border with a more strengthened presence, as states called for additional deployments. Their relevance indicates that Warsaw has not yet become obsolete.

Conclusion

Both borders and culture are not fixed. They are dynamic, they can change, and constantly face novel challenges. In fact, as this book was nearing completion, EU borders became shaken by the outbreak of war in a neighbouring country. Upon their emergence, new challenges can knock over established border control routines or contest taken-for-granted border assumptions. Put differently, they can challenge Warsaw culture and Frontex's dominant position in the border control community, as they signal and can provoke fundamental shifts both in the environment and the actors of border control.

On that account, the chapter sought to reflect on certain major challenges facing Warsaw culture and the border control policy community. These are Brexit and the English Channel's transformation from internal to external border, the COVID-19 pandemic as well as hard security preoccupations triggered by the Greek-Turkish stand-off in Evros and the war in Ukraine. Zooming into the function of borders and the border control conduct during these events, the chapter produced a puzzling research result, that is, the existence of elements conjointly contradicting and complying to Warsaw culture. Two of those elements include militarisation and multilateral cooperation in the case of the 2020 Evros crisis. Militarisation is not part of the Warsaw trajectory; in contrast, multilateral cooperation constitutes one of Warsaw's border control practices. The pursuit and adoption of border control elements foreign to Warsaw, such as militarisation, signify that this culture has started to be challenged. Yet, despite being challenged, it has not been replaced by another culture of border control. It retains its dominant position in EU border control, taking into account that Warsaw's tenets, like multilateral cooperation, still mould the EU borders. In addition, Warsaw's effective contestation becomes highly unlikely considering the lack of a common approach when addressing the new developments at the EU borders. Accordingly, amid the COVID-19 pandemic all borders closed, restricting mobility, whereas the war in Ukraine brought the abolishment of all border restrictions. The divergence in attitudes undermines the prospect of Warsaw's replacement or, put differently, evolution, as there is no consistency or united front among the alternative border control paradigms. So, the cultural evolution process starting with Warsaw's variation has not yet taken place.

As it is drawn from the cultural evolution process and mechanism (Zaiotti, 2011), challenges to the dominant culture of border control are expected to occur both in the long run and short run. In that spirit, it is not surprising to draw elements at the borders that do not belong to Warsaw culture; especially taking into consideration the magnitude of the developments currently taking place at the borders, such as the outbreak of war in Europe and a global health alert. In this new environment and amid facing novel challenges or crises, priorities and imperatives are to be readjusted. Whether these new elements will convert from a temporary toolkit to the new cultural norm remains to be seen. After all, cultural change, or put differently, cultural evolution, cannot occur imminently (Meyer, 2006: 15–42). It is a long process unfolding over time and a demanding task involving different steps taking place gradually (see Chapters 1 and 6). So, Warsaw's fate cannot be currently forejudged.

The same applies to Frontex. It is too early to determine how these developments will affect Frontex's future. Irrespective of Frontex's key position in Warsaw's border control community as well as the agency's role in Warsaw's promotion illustrated throughout this book, Frontex is not synonymous to

Warsaw culture. The agency existed well before Warsaw culture. So, it may continue existing after Warsaw's dissolution and replacement by another culture of border control. After all, Frontex, being an actor, has a primary goal, that is, to ensure its survival by continuing being relevant. To stay relevant, Frontex became actively involved at the EU borders during all the new challenges discussed in this chapter, regardless of their nature. It deployed resources at the UK-France border after Brexit, having as its aim to prevent the outflow of irregular border crossers, although Frontex's typical 'business' concerns the combat of irregular entries to the EU. Amid the COVID-19 pandemic outbreak and in defiance of its mandate lacking any health reference, Frontex claimed and succeeded in expanding its role and scope of action by starting to cover protection from health threats. Even during hard security considerations, like in Evros or the war in Ukraine, Frontex did not just continue being present at the borders. It expanded its operational role by conducting new missions and bringing more Frontex officers on the ground. Hence, Frontex met all the arising challenges by converting them into opportunities for the agency's role enhancement.

Looking ahead, to retain its comparative advantage and maintain being perceived as a valuable asset in EU border control, Frontex must continue confronting head on all the challenges, either internal or external, as well as keep up with new trends, at global, EU, or national level, even if that means a move towards a wider border lexicon. After all, as this chapter illustrated, Frontex's widening has already been set in motion. So, now Frontex presides over its own fate. Whether it will continue to rule over the EU border control or become obsolete is still open for discussion.

Notes

1 UK entered the European Communities in 1973.
2 After the Council's rejection of the UK's demand to participate in Frontex, both the UK and Northern Ireland brought the case to the European Court of Justice seeking the annulment of Council's decision. The Court dismissed the case by concluding that Frontex's functions were 'constituting elements of the Schengen acquis' (CJEU, 2007).
3 In May 2022, 2,871 irregular border crossers on small boats were recorder compared with 1,627 in May 2021 (The Guardian, 2022).
4 Due to its geographic vicinity, Belgium has occasionally participated in discussions with the UK and France on migration flows in the English Channel.
5 Until April 2022, more than 510,270,660 cases of COVID-19 have been confirmed incurring approximately 6,233,520 deaths. From the cases reported, 214,635,881 referred to Europe (WHO, 2022b).
6 The 4 Presidents that visited Evros were the President of the European Council, Charles Michel, the President of the European Commission, Ursula von der Leyen, the President of the European Parliament, David Sassoli, and the Prime Minister of Croatia, Andrej Plenković, whose country was holding the rotating Presidency of the Council of the European Union at the time.
7 The Turkish National Assembly adopted a resolution on 8 June 1995 granting the Turkish government full and perpetual competence to declare war (casus

belli), should Greece decide to extend its territorial waters over 6 nautical miles.
8 For more information on the Maidan protests, see Shore (2017).
9 At the time of writing these pages (April-May 2022), the war is still waging.
10 Hungary was an exception (ECRE, 2022).

References

BBC (2015) 'Who, What, Why: What exactly is the UK's National Barrier Asset?'. Available at https://www.bbc.com/news/magazine-33316358 (accessed April 2022).
BBC (2016) 'Calais Migrants: How is the UK-France Border Policed?'. Available at https://www.bbc.com/news/uk-33267137 (accessed April 2022).
Berger, T.U. (1996) 'Norms, Identity and National Security in Germany and Japan'. In: P.J. Katzenstein (ed.), *The Culture of National Security: Norms and Identity in World Politics*. New York: Columbia University Press, pp. 317–356.
Buller, J. (1995) 'Britain as an Awkward Partner: Reassessing Britain's Relations with the EU'. *Politics*, 15(1), pp. 33–42.
CJEU (2007) United Kingdom v Council. Case C-77/05.
CoE (2020) 'Council of Europe's Anti-Torture Committee Calls on Greece to Reform its Immigration Detention System and Stop Pushbacks'. Available at https://www.coe.int/en/web/cpt/-/council-of-europe-s-anti-torture-committee-calls-on-greece-to-reform-its-immigration-detention-system-and-stop-pushbacks (accessed April 2022).
Commission of the European Communities (2003) 'Proposal for a Council Regulation establishing a European Agency for the Management of Operational Co-Operation at the External Borders'. COM(2003)687.
ComputerWeekly (2022) 'English Channel Surveillance Used to Deter and Punish Migrants'. Available at https://www.computerweekly.com/feature/English-Channel-surveillance-used-to-deter-and-punish-migrants (accessed April 2022).
Council Directive 2001/55/EC of 20 July 2001 on Minimum Standards for Giving Temporary Protection in the Event of a Mass Influx of Displaced Persons and on Measures Promoting a Balance of Efforts between Member States in Receiving such Persons and Bearing the Consequences Thereof [2001, OJ L 212/12].
Davies, T., Isakjee, A., Mayblin, L. & Turner, J. (2021) 'Channel Crossings: Offshoring Asylum and the Afterlife of Empire in the Dover Strait'. *Ethnic and Racial Studies*, 44(13), pp. 2307–2327.
ECRE (2022) 'Ukrainian Borders'. Available at https://ecre.org/ukrainian-borders-more-than-two-million-arrivals-to-neighbouring-states-strong-community-response-in-poland-hungary-introducing-tougher-controls-moldova-under-pressure/ (accessed April 2022).
Ekathimerini (2021) 'We Fought Man-to-Man to Hold the Evros Border'. Available https://www.ekathimerini.com/in-depth/special-report/1156395/we-fought-man-to-man-to-hold-the-evros-border/ (accessed April 2022).
Eržen, B., Weber, M. & Sacchetti, S. (2020) 'How COVID-19 is Changing Border Control'. ICMPD, Expert Voice. Available at https://www.icmpd.org/news/how-covid-19-is-changing-border-control (accessed April 2022).
EURACTIV (2020) 'Erdogan Vows to Keep Border Open, Compares Greece's Response to the Nazis'. Available at https://www.euractiv.com/section/justice-home-affairs/news/erdogan-vows-to-keep-border-open-compares-greeces-response-to-the-nazi/ (accessed April 2022).

Euronews (2021) 'Frontex Plane Arrives in Northern France to Help Fight People Smuggling'. Available at https://www.euronews.com/2021/12/02/frontex-plane-arrives-in-northern-france-to-help-fight-people-smuggling (accessed April 2022).

European Commission (2020a) COVID-19: Guidelines for Border Management Measures to Protect Health and Ensure the Availability of Goods and Essential Services [2020, OJ C 86 I/1].

European Commission (2020b) 'Commission Staff Working Document: Turkey 2020 Report'. SWD(2020)355.

European Commission (2022) 'Travel during the Coronavirus Pandemic'. Available at https://ec.europa.eu/info/live-work-travel-eu/coronavirus-response/travel-during-coronavirus-pandemic_en (accessed April 2022).

Foreign Affairs (2021) 'How Migrants Got Weaponized'. Available at https://www.foreignaffairs.com/articles/2021-12-02/how-migrants-got-weaponized (accessed April 2022).

Foucault, M. (2008) 'The Birth of Biopolitics: Lectures at the Collège de France 1978-1979'. In: M. Senellart (ed.), *Michel Foucault*. Basingstoke: Palgrave Macmillan.

Frontex (2006) *Frontex Annual Report 2006*. Warsaw: Frontex.

Frontex (2010) *Frontex General Report 2010*. Warsaw: Frontex.

Frontex (2014) 'Joint Return Operation RO to Nigeria by United Kingdom on 28.01.2014'. *Frontex Evaluation Report*. Warsaw: Frontex.

Frontex (2016) *Risk Analysis for 2016*. Warsaw: Risk Analysis Unit.

Frontex (2020a) *Risk Analysis for 2020*. Warsaw: Risk Analysis Unit.

Frontex (2020b) 'Europe Day-United against Corona Virus with Eyes on the Future'. Available at https://frontex.europa.eu/media-centre/news/news-release/europe-day-united-against-corona-virus-with-eyes-on-the-future-r9vMlS (accessed April 2022).

Frontex (2021a) *Risk Analysis for 2021*. Warsaw: Frontex.

Frontex (2021b) *Annual Implementation Report 2020*. Warsaw: Frontex.

Frontex (2021c) 'Frontex to Support Member States in the Channel and North Sea Region'. Available at https://frontex.europa.eu/media-centre/news/news-release/frontex-to-support-member-states-in-the-channel-and-north-sea-region-pZWNYE (accessed April 2022).

Frontex (2021d) *Single Programming Document 2022-2024*. Warsaw: Frontex.

Frontex (2022a) 'Who We Are'. Available at https://frontex.europa.eu/about-frontex/who-we-are/foreword/ (accessed April 2022).

Frontex (2022b) 'Frontex Deploys its Own Plane in the Channel'. Available at https://frontex.europa.eu/media-centre/news/news-release/frontex-deploys-its-own-plane-in-the-channel-cHkg9q (accessed April 2022).

Frontex (2022c) 'Frontex Ready to Support Member States in light of Situation in Ukraine'. Available at https://frontex.europa.eu/media-centre/news/news-release/frontex-ready-to-support-member-states-in-light-of-situation-in-ukraine-kZGGwq (accessed April 2022).

Frontex (2022d) 'On Europe Day We Stand with Ukraine' [@Frontex]. Tweet available from https://twitter.com/frontex?lang=el (accessed May 2022).

Goodman, S. (2017) 'Take Back Control of Our Borders: The Role of Arguments About Controlling Immigration in the Brexit Debate'. *Yearbook of the Institute of East-Central Europe*, 15(3), pp. 35–53.

Greek Council for Refugees (2021) 'Country Report: Reception and Identification Procedure'. Available at https://asylumineurope.org/reports/country/greece/asylum-procedure/access-procedure-and-registration/reception-and-identification-procedure/ (accessed April 2022).

Greek PM (2020) 'Statement by Prime Minister Kyriakos Mitsotakis in Kastanies'. Available at https://primeminister.gr/en/2020/03/03/23458 (accessed April 2022).

HM Government (2013) 'Decision Pursuant to Article 10 of Protocol 36 to the Treaty on the Functioning of the European Union'. Cm 8671.

House of Commons (2019) 'Migrants Crossing the English Channel'. Available at https://commonslibrary.parliament.uk/migrants-crossing-the-english-channel/ (accessed April 2022).

House of Commons (2022) 'Operation Isotrope: The Use of the Military to Counter Migrant Crossings'. *Defence Committee*, HC 1069.

Kirk, J. & McDonald, M. (2021) 'The Politics of Exceptionalism: Securitization and COVID-19'. *Global Studies Quarterly*, 1(3), pp. 1–12.

Kostas, S. (2017) 'Irregular vs. Illegal Immigration: Setting the Definitions. An Overview of European Practice'. *Slovenský Národopis*, 65(4), pp. 420–442.

Lewens, T. (2015) *Cultural Evolution: Conceptual Challenges*. Oxford: Oxford University Press.

Management Board (2016) Frontex Management Board Decision No 45/2016 on the United Kingdom's Participation in Frontex Joint Operation EPN Triton 2016 and Joint Operation EPN Poseidon 2016.

Management Board (2021) Frontex Management Board Decision 36/2021 of 16 June 2021 Authorising the Executive Director to Negotiate Working Arrangements with Selected Third Countries.

Martin, J. (1992) *Cultures in Organizations: Three Perspectives*. Oxford: Oxford University Press.

Meyer, C.O. (2006) *The Quest for a European Strategic Culture: Changing Norms on Security and Defence in the European Union*. Basingstoke: Palgrave Macmillan.

Mitsilegas, V. (2017) 'European Criminal Law after Brexit'. *Criminal Law Forum*, 28(2), pp. 219–250.

Morieux, R. (2016) *The Channel: England, France and the Construction of a Maritime Border in the Eighteenth Century*. Cambridge: Cambridge University Press.

OHCHR (2022) 'Ukraine: Civilian Casualty Update 2 May 2022'. Available at https://www.ohchr.org/en/news/2022/05/ukraine-civilian-casualty-update-2-may-2022 (accessed May 2022).

Parker, S., Bennett, S., Cobden, C.M. & Earnshaw, D. (2021) 'It's Time We Invested in Stronger Borders: Media Representations of Refugees Crossing the English Channel by Boat'. *Critical Discourse Studies*. https://doi.org/10.1080/17405904.2021.1920998 (published online).

Politico (2020) 'Greece Says it will Stop Accepting Asylum Requests amid Migrant Crisis'. Available at https://www.politico.eu/article/frontex-sends-reinforcements-greece-migrant-crisis-turkey/ (accessed April 2022).

Politico (2021) 'EU Approves Mandate for Gibraltar Treaty Negotiations with UK'. Available at https://www.politico.eu/article/eu-mandate-gibraltar-treaty-negotiations-uk/ (accessed April 2022).

Regulation (EC) No 2007/2004 of 26 October 2004 Establishing a European Agency for the Management of Operational Cooperation at the External Borders of the Member States of the European Union [2004, OJ L 349/1].

Regulation (EC) No 562/2006 of the European Parliament and of the Council of 15 March 2006 Establishing a Community Code on the Rules Governing the Movement of Persons across Borders (Schengen Borders Code) [2006, OJ L 105/1].

Regulation (EU) 2016/399 of the European Parliament and of the Council of 9 March 2016 on a Union Code on the Rules Governing the Movement of Persons across Borders (Schengen Borders Code) [2016, OJ L 77/1].

Regulation (EU) 2019/1896 of the European Parliament and of the Council of 13 November 2019 on the European Border and Coast Guard and Repealing Regulations (EU) No 1052/2013 and (EU) 2016/1624 [2019, OJ L 295/1].

Scott, W.R. (2008) *Institutions and Organizations: Ideas and Interests.* London: SAGE.

Shore, M. (2017) *The Ukrainian Night: An Intimate History of Revolution.* Haven: Yale University Press.

Swidler, A. (1986) 'Culture in Action: Symbols and Strategies'. *American Sociological Review,* 51(2), pp. 273–286.

The Guardian (2015) 'The Horror of the Calais Refugee Camp'. Available at https://www.theguardian.com/world/2015/nov/03/refugees-horror-calais-jungle-refugee-camp-feel-like-dying-slowly (accessed April 2022).

The Guardian (2020) 'EU and Turkey Hold "Frank" Talks over Border Opening for Refugees'. Available at https://www.theguardian.com/world/2020/mar/09/turkey-erdogan-holds-talks-with-eu-leaders-over-border-opening (accessed April 2022).

The Guardian (2021) 'Channel Drownings: UK and France Trade Accusations after Tragedy at Sea'. Available at https://www.theguardian.com/uk-news/2021/nov/25/channel-drownings-uk-and-france-trade-accusations-after-tragedy-at-sea (accessed April 2022).

The Guardian (2022) 'Priti Patel's Plan to End Channel Crossings in Disarray as Navy Threatens to Walk Away'. Available at https://www.theguardian.com/world/2022/jul/09/priti-patels-plan-to-end-channel-crossings-in-disarray-as-navy-threatens-to-walk-away (accessed July 2022).

The Times (2021) 'France Blocks British Company from Patrolling for Migrants'. Available at https://www.thetimes.co.uk/article/france-blocks-british-company-from-patrolling-for-migrants-w9w2d35dr (accessed April 2022).

Treaty on the Functioning of the European Union (TFEU) (2016) Protocol (No 20) on the Application of Certain Aspects of Article 26 of the Treaty on the Functioning of the European Union to the United Kingdom and to Ireland [2016, OJ C 202].

UK Foreign Secretary (2021) 'A Treaty between the UK and EU in respect of Gibraltar: Joint Ministerial Statement'. Available at https://www.gov.uk/government/news/joint-ministerial-statement-on-a-treaty-between-the-uk-and-eu-in-respect-of-gibraltar (accessed April 2022).

UK Home Office (2015) 'Joint UK/French Ministerial Declaration on Calais'. Available at https://www.gov.uk/government/publications/joint-ukfrench-ministerial-declaration-on-calais (accessed April 2022).

UK Home Office (2018) 'Joint UK-France Centre Opens in Calais to Tackle Criminality at Border'. Available at https://www.gov.uk/government/news/joint-uk-france-centre-opens-in-calais-to-tackle-criminality-at-border (accessed April 2022).

UK Home Office (2019) 'UK-France Joint Action Plan on Illegal Migration across the Channel'. Available at https://www.gov.uk/government/publications/uk-france-joint-action-plan-on-illegal-migration-across-the-channel (accessed April 2022).

UK Home Office (2020a) 'Media Factsheet: Small Boats'. Available at https://homeofficemedia.blog.gov.uk/2020/05/15/media-factsheet-small-boats-2/ (accessed April 2022).

UK Home Office (2020b) 'Home Secretary Appoints Small Boat Commander'. Available at https://www.gov.uk/government/news/home-secretary-appoints-small-boat-commander (accessed April 2022).

UK Home Office (2022) 'Official Statistics: Irregular Migration to the UK'. Available at https://www.gov.uk/government/statistics/irregular-migration-to-the-uk-year-ending-december-2021/irregular-migration-to-the-uk-year-ending-december-2021 (accessed April 2022).

UN (2014) 'Backing Ukraine's Territorial Integrity: UN Assembly Declares Crimea Referendum Invalid'. Available at https://news.un.org/en/story/2014/03/464812-backing-ukraines-territorial-integrity-un-assembly-declares-crimea-referendum (accessed April 2022).

UNHCR (2022a) 'Mediterranean Situation'. Available at https://data2.unhcr.org/en/situations/mediterranean (accessed April 2022).

UNHCR (2022b) 'Syria Regional Refugee Response: Turkey'. Available at https://data2.unhcr.org/en/situations/syria/location/113 (accessed April 2022).

UNHCR (2022c) 'Ukraine Refugee Situation'. Available at https://data2.unhcr.org/en/situations/ukraine (accessed April 2022).

van Eijken, H. & Rijpma, J. (2021) 'Stopping a Virus from Moving Freely: Border Controls and Travel Restrictions in Times of Corona'. *Utrecht Law Review*, 17(3), pp. 34–50.

VOA (2021) 'Smugglers Net Millions per Kilometer from Migrants Crossing Channel'. Available at https://www.voanews.com/a/smugglers-net-millions-per-kilometer-from-migrants-crossing-channel/6330386.html (accessed April 2022).

WHO (2022a) 'Coronavirus Disease (COVID-19) Pandemic'. Available at https://www.euro.who.int/en/health-topics/health-emergencies/coronavirus-covid-19/novel-coronavirus-2019-ncov (accessed April 2022).

WHO (2022b) 'WHO Coronavirus (COVID-19) Dashboard'. Available at https://covid19.who.int/ (accessed April 2022).

Zaiotti, R. (2011) *Cultures of Border Control: Schengen & the Evolution of European Frontiers*. Chicago: University of Chicago Press.

8 Conclusion

Frontex's leadership and the re-drawing of EU border control

> 'This unique and undisputable way [...] identifies the work of Frontex [...], the Frontex way'.
> —Fabrice Leggeri, former Frontex Executive Director
> (Frontex, 2021: 5)

Amid turbulent or even shocking times that provoke disorder and test the EU's unity, the EU borders are being redrawn. Since Croatia's accession in 2013 and the UK's withdrawal in 2020 that resketched the EU map, several applications for EU membership are being processed; some recent, others older. Notwithstanding the application date, their approval may lead to an eventual expansion of the EU geographic area. In fact, the accession of new member countries to the EU will shift borders, both internal and external, reshaping therefore the EU territory and border construction. Until then, though, the EU borders remain largely unquestioned, enabling the actors that manage them to continue unhindered ordering flows, regular or irregular, of people and goods according to established processes as well as taken for granted border understandings, as cross-border mobility never ceases.

Irrespective of any expansion or subtraction of the EU geographic area, borders keep being on the mind (Agnew, 2008). Without receding, the prevalence of old and emerging border anxieties reflects their never-decreasing relevance for both the EU project and its member states. Endlessly, the erection of borders, both physical and mental, functions as a means to fence off interests, territorial integrity, and the (EU)ropean identity by keeping the 'outsiders' and those being portrayed as the 'others' away, reinvigorating new dividing lines in an attempt to safeguard the 'European way of life'. In this endeavour, the institutions that manage the borders acquire an expanded role rendering the control of the border a task of pivotal importance for the salvation of the national security and the EU order. Such an institution is Frontex, the EU border control agency, engaged with EU border control since its operational launch in 2005.

DOI: 10.4324/9781003230250-8

176 *Conclusion*

In fact, besides national border guards, Frontex has been present at the border since 2005. Either with border officers deployed at the borders in the framework of a Frontex operation or through other Frontex activities, such as data collection and risk analysis performed by its staff at Frontex's headquarters. So, Frontex has evolved into an indispensable entity at EU borders and for EU border control. Inspired by that, this book was written with a fairly straightforward aspiration: to explore Frontex's role in EU border control. I attempted to highlight that Frontex is not a simple extension of member states' remit. Rather, it constitutes a key border control actor that rules over the nature and direction of EU border control by shaping its cultural disposition.

In this concluding chapter, I provide an overview of the book's main findings, namely the emergence of a new dominant culture for border control in Europe and Frontex's cultural impact on EU border control. Furthermore, I revisit the conclusions drawn about Frontex, borders, and EU border control. Lastly, reaching to the end of this book's exploration of Frontex's role in EU border control, I reflect on the themes and debates that have surrounded the book by offering some closing thoughts on borders and Frontex's role in EU border control, including present and future directions.

Evolving borders

The rationale of this book was that borders are more than an image or a cartographic representation. They are social constructions (Newman & Paasi, 1998: 187; Newman, 2006: 173), which are being made (van Houtum et al., 2005; Newman, 2011). They are being made by the actors that manage them, namely the institutional apparatus that governs borders, and the rationalities that define their functions delineating the insiders from the outsiders. From that vantage point, every border is different. The Greek-Turkish land border varies from the Greek-Albanian border. Equally different are internal borders; for instance, Austria's border with Germany differs from the Austro-Hungarian border.

Besides the geographic determinants causing variation, there are dissimilarities encountered in the procedures to enter a country, the intensity of border control, and the disposition of border officers upon certain nationalities. Accordingly, on certain intra-Schengen flights departing from Greece to northern Europe destinations, such as France, Luxembourg, and Germany, during the embarkation to the airplane, travellers not looking like 'Europeans' are often being set aside by employees of the airline company, so that police officers can check their travel documents more thoroughly. Even the gender, age, appearance, travel companions, and behaviour of the traveller can lead to stricter or lenient border checks. For instance, in a recent trip from France to the USA (in May 2022), I witnessed French border officers conducting 'random' luggage control, mostly on male passengers travelling alone, even at the gate, that is, just before entering the airplane.

In that context, both borders and border control are not static or fixed. They are not monolithic. They shift, revealing inter-subjective interferences (Bellamy et al., 2017: 484). At the same time, borders produce symbolic meanings by regulating practices of power (Newman, 2003; van Houtum et al., 2005), such as functional or symbolic rationalities of inclusion and exclusion (Diez et al., 2006). In that regard, border guards have the power to decide who will be controlled as well as who belongs inside the territory, who can enter, and who will remain outside of it. So, borders evolve into a 'field of action' and a 'basis for action' (Lefebvre, 1991: 191), among those controlling the border and those trying to cross it, regularly or irregularly.

The above manifest that borders enclose a dynamic nature. They can change. Equally, border control is not static but can evolve and adapt. It can be conducted by new border control actors. It can be produced by a different border control community. It can be shaped by a new set of border control assumptions and practices. In other words, it can follow a different cultural trajectory, namely a new culture of border control. This realisation enabled a more in-depth analysis of the structure and conduct of border control. Shedding light upon the underlying social dynamics within the border control community, the book revealed both the rise of a new border control culture and the true role of the most prominent of the border control actors, that is, Frontex.

The rise of a new border control culture

To explore Frontex's role in EU border control, the book zoomed into Frontex and the EU borders. Accordingly, drawing on culture and applying the cultures of border control analytical framework (Zaiotti, 2011), it investigated, first, an institutional actor of border control, namely Frontex (Chapters 2 and 3), and, then, two different EU external borders, namely the Italian sea border of Lampedusa (Chapter 4) and the Greek land border of Evros (Chapter 5). The analysis unveiled the emergence of a new culture of border control in Europe, which this book named 'Warsaw culture'.

The discovery of certain common assumptions and practices for the border control conduct, not only in different border locations but also within a border control actor, Frontex, attests that they are not isolated characteristics emerging in a particular geographic area or institutional context. Instead, they are components of a border control culture currently pursued.

Unravelling this culture and naming it as 'Warsaw culture', the book, adopting a comparative disposition, showed that it constitutes a new and unique cultural trajectory not linked to the other cultures of border control in Europe, namely 'Schengen', 'Westphalia', and 'Brussels' (Chapter 6). In fact, following the steps of the cultural evolution process for Warsaw's institutionalisation, the analysis ascertained that Warsaw culture has now become the dominant border control culture after replacing Schengen culture. Hence, a new border control culture has not only risen in Europe, but most importantly, it has evolved into a dominant cultural trajectory for border control, fundamentally shaping the EU borders.

The shared border control assumptions traced, which compose Warsaw culture, include securitisation, technocracy, surveillance, intelligence, and a re-territorialisation of the place of the border control conduct. The common border control practices involve information gathering, multilateral cooperation, technology, and professionalisation. Moreover, on border assumptions, Warsaw culture, adopting a binary conception of borders, perceives them as both national and EU external borders.

Along with the discovery of Warsaw culture and its constitutive assumptions and practices, the book also attested the consolidation of a border control community, which operates at the borders, shaping the border control conduct. This community has been labelled as 'practitiocratic', because it is being composed of border control practitioners, namely border guards and Frontex staff. This community has been formed as a result of the interaction of border control practitioners. In particular, it was formed when border control practitioners from different EU member states started to work together for the management of the EU borders, developing social relations as well as sharing interests, preoccupations, and incentives. Coming into direct contact and building social relations they acknowledged their common disposition, which led them to gradually function as a community and then pursue a new cultural trajectory for border control. Actually, border control practitioners were not involved in the design and development of Schengen culture, which constituted the then-dominant culture of border control in Europe. Their omission from Schengen culture led them, upon their consolidation as a border control community, to select and pursue a different cultural trajectory for border control, that is, Warsaw culture.

Bringing the cultures of border control analytical framework into Frontex's analysis, this book provided a platform for the framework's application. Since its initial conception and application by Zaiotti, who studied Schengen's emergence in the mid-1980s and 1990s, it had never been used as a main analytical tool in further cases.[1] Thus, its inclusion here proves — generally — the framework's validity and explanatory strength, attesting that it can be employed as an analytical tool in other studies about border control. Moreover, given that Zaiotti's analysis was conducted in a previous chronological period, that is, more than 10 years ago, this research, by exploring the current border control conduct, brings new findings for today's borders and border control in Europe. In parallel, it attests the framework's continued relevance for the analysis of EU border control.

The Frontex effect

Analysing the EU borders and the border control conduct, the book also assessed Frontex's effect on EU border control. More specifically, it unfolded Frontex's distinctive role in reconfiguring border control via the development and promotion of a new culture of border control.

Conclusion 179

17 years after Frontex's operational launch, this book proposed to rethink Frontex's function and role at Europe's borders, adopting a fresh prism and moving beyond any theoretical and methodological boundaries set by pre-existing framings of Frontex. Actually, from day one of its establishment, Frontex has attracted considerable attention from the media, the public, and academia, engaging in intensive discussions about its function, raising human rights considerations, critiquing its security context as well as highlighting continuous institutional enhancements. While those aspects are important in addressing critical aspects of Frontex, at the same time they constituted the sole basis of any preoccupation with this EU agency, impeding therefore the exploration of other dimensions of Frontex's function. Following that, the attempt to distance from them by pursuing a different research path was a challenge.

Another challenge stemmed from Frontex's conception as an instrument of border control. Drawing on this prevailing opinion, one possibly would have thought that Frontex is not that important. It is just an EU agency, namely a pawn of member states or of the EU Commission that through the creation of Frontex found a way to bend member states' objections and become involved in the sphere of border control. So, what is the point in studying an EU agency?

Overcoming these challenges, I hope that this book has shown throughout its pages that Frontex is still relevant and important, with certain veiled aspects regarding its role. In fact, it is an actor and not an instrument of border control, as the conventional wisdom on Frontex usually holds. Being an actor, it is capable of producing new meanings and actions as well as promoting its own assumptions and practices for border control (Chapter 3). In that regard, Frontex is an actor that impacts fundamentally the EU border control. Its impact refers to the emergence, (re)production, and promotion of Warsaw culture.

Indeed, Frontex has led to a notable change in EU border control. That change is not limited to Frontex's presence at the borders by deploying Frontex guest officers and staff from its standing corps. Rather, the book demonstrated that Frontex led to a variation in the culture of border control pursued. In fact, Frontex has produced Warsaw's border control community, reference texts, as well as border control assumptions and practices. So, Frontex does not just function as an instrument of EU border control. Rather, it reconfigures border control through shaping its culture.

Before Frontex, there was a variant border control trajectory. Frontex, not being part of this structure, promoted a different regime, solidifying, in turn, its position in border control (Chapter 6). That refers to a rational decision and a strategic objective pursued by Frontex so as to strengthen its role and ensure its institutional existence. Being the one that actively develops and promotes the culture that has become the dominant culture of border control after replacing Schengen culture, it keeps aside potential 'adversaries', namely other actors aspiring to rule the EU borders, maintaining its central role in EU border control. Thus, Frontex shifted the old structure and initiated a new one, implemented with the promotion of a variant culture of border control.

But what does the assertion that border control is being impacted by Frontex really mean? A preliminary answer would be that Frontex is not an accessory of EU border control. Although it has been created to respond to border control needs, functioning as a tool, Frontex has now evolved into an agent or architect of EU border control, having the ability to form its own border control imperatives and promote them next to others. In other words, rather than functioning in line with pre-established border control meanings, it creates its own meanings, which are then diffused to others. The meanings that it deems more appropriate or best suited to its own needs. So, instead of conforming to the border control context that was formed before Frontex, Frontex managed to co-form it by promoting a new border control trajectory.

Yet, Frontex has not been created to shape or co-shape EU border control but just to assist member states with the implementation of external border management (Regulation, 2004). Neither was it anticipated that in the long run Frontex could bring such a fundamental impact through its engagement with border control.

Border control is a policy field affecting both the EU citizens and third country nationals crossing the border, either regularly or irregularly. On that account, Frontex's impact on border control means that it should be made accountable to them. But, until now, as Chapter 2 noted, despite control mechanisms on Frontex, such as through the control of budget allocation or fundamental rights monitoring, the agency still savours autonomy even on technical issues. Let alone on issues not anticipated by its creators that the agency would engage with and assume an activity. Actually, Frontex is still not perceived by the academia and national or EU decision-makers as an agent capable of producing a cultural impact on the EU border control. Drawing on that, there is no plan to control such impact or restrict it. As a result, Frontex is free to act as it deems most suitable to its own needs.

Besides border control, or better put, through its engagement with border control, Frontex also impacts borders. As already described whilst analysing the cultural traits of Frontex as well as Warsaw culture, Frontex embodies certain assumptions about borders, which, though they are not a novelty or absurd, have still been transfigured into border assumptions, acquiring therefore a cultural disposition via Frontex's intervention. That means that Frontex, by solidifying its position in EU border control and consolidating its agency through the promotion of Warsaw culture, has also constructed territoriality (Della Sala, 2017: 546), initiating its own representations and meanings about borders. So, Frontex has reconfigured not only border control and its culture but also borders.

Acknowledging the role of Frontex upon the reconfiguration of the EU border control and the EU borders opens up the floor for deliberations about Frontex's power. Indeed, the management of borders reconstructs different forms of power. Therefore, borders 'are used by various bodies and institutions in the perpetual process of reproducing [power]' (Paasi, 2009: 213), such as territorial, symbolic, social, institutional, and functional power. In that spirit, any institution being at the borders and managing borders unavoidably becomes involved in the reproduction of forms of power. The extent, shape, and kind of power may vary in relation to the institution's role in the border management.

Although I approached Frontex without referring to power, 'hard' or even 'soft' power, inserting instead a cultural lens, the findings drawn from the scrutiny of Frontex's role in EU border control imply that Frontex is not powerless. Its power, though, is different. It stems from its central role in the border control community, its monopoly over the EU border control as the sole actor at the EU level participating in border management, and its contribution in the development of the dominant culture of border control shaping Europe's borders.

Frontex's cultural impact

Studying borders and Frontex, the book offers a different understanding regarding the drivers, dynamics, and evolution of border control in Europe. It captures that a border control actor, namely Frontex, can impact borders and border control through culture. It should be noted that in the book I do not claim that Frontex is the sole border control actor. Neither that Frontex is the only producer and promoter of cultural norms. But I do hope to have shown that any study of the EU borders that disregards Frontex seems lacking, as Frontex has now evolved into a fundamental aspect of the EU border control and an essential border control actor at Europe's borders. At the same time, the inclusion of culture in an analysis of Frontex can illuminate important dimensions regarding the agency's function and role in EU border control.

It follows from the foregoing that decision-making is not the only path for exercising an impact on this policy domain. Apart from decision-makers, Frontex, though officially remaining outside of the EU decision-making mechanism, can have — and as shown in this book has — a pivotal effect on EU border control. Besides its role in the implementation of border control policies through the tasks delegated to this EU agency by the member states and the European Commission or its power to inform decisions through the provision of data, technical knowledge, and expertise, Frontex also informs border control via culture. Hence, Frontex's impact on EU border control is not one-dimensional. Rather, it is linked to its multifaceted role and therefore it concerns a wider and diverse context. If one were using a conceptual scale, then one edge would refer to the technical parts of Frontex's function and impact, such as the provision of factual evidence or statistical data; the middle ground would have been occupied by Frontex's policy intervention performed, for example, through the articulation of strategic risk analysis, border management strategies, conduct of joint missions, and evaluation of policy implementation; whereas the other edge would involve Frontex's cultural disposition with the development and promotion of Warsaw culture.

Needless to say, all parts of Frontex's impact are equally important, necessary, and maybe a little unexpected, given that we still talk about an EU agency treated mostly as an instrument and not an actor. Yet, the cultural aspect seems the most intriguing as it refers to a subtle and gradual impact, requiring both a considerable and coordinated effort involving the internalisation of cultural traits and then their transmission to others; with dubious results as no one can be sure in advance about the success of such an endeavour and whether it would finally lead to culture's retention and consolidation as the dominant culture, replacing the other cultural trajectories. In spite of its ambiguity, such vigour can lead to a profound shift in EU border control, that is, the adoption of a different cultural trajectory. So, speaking of culture, a cultural impact is far more substantive than any other change triggered through alternative channels. It is all-encompassing and therefore it has the ability to shake deep-rooted conceptions as well as deem previously accepted modes of action as obsolete. After all, it is culture that makes some things possible, while others unimaginable (Kier, 1997: 65). It is culture that rules sentiments, judgments, and actions (Gray, 1999: 143). It is culture that guides thinking, feeling, and acting (Alvesson, 2013: 6). In that spirit, a cultural impact overrules all the others.

Following that, a question arises: what does it mean for border control to be culturally impacted and -shaped? Border control is usually acknowledged and treated as a highly technical field characterised by professional jargon and the deployment of specialised border control tools. Yet, the condition of culture renders border control as a 'playing field' for new considerations. Accordingly, border control being informed by culture or shaped according to cultural norms as well as border control actors being embedded in a cultural setting presupposes a common way of thinking about borders and border control (Meyer, 2006: 3). That makes clear that the border control tactics or tools employed are not random. Neither do they constitute part of an occasional arrangement. Nor do they come up in the wake of a random crisis or imminent threat at the borders. Instead, they compose culture-conditioned responses drawn from a pre-formed and pre-disposed repertoire. Put differently, they reflect behavioural choices being shaped by culture. From that perspective, border control transcends its technocratic nature. It adopts a deeper and more complex foundation, being informed by collective meanings and shared understandings. Indeed, it becomes linked to non-material dispositions about the role of the borders and the importance of border control, as culture takes into account ideational conditions; it encloses dynamic characteristics, as culture can change; and lastly it displays deep-seated beliefs, as culture stabilises inter-subjective understandings. So, border control is more than technical. It is a cultural expression. Therefore, it mirrors values and perceptions as well as defines threat assessments, goals, and roles attributed to border control actors designating 'what is right', 'what works' in EU border control, 'for what purpose', and 'with what tools'.

A cultural configuration of border control does not only refer to present or past policy choices. It also addresses its future development as it provides an outlook of its prospective direction. Likewise, it forecasts borders' function, given that border control is inextricably linked with borders. That gains added weight in the context of an increasing cross-border mobility in the not too far distant future. According to Frontex's projections, the EU will become the main destination for both migrants and asylum seekers because of the high population growth in Sub-Saharan Africa and Western Asia (Frontex, 2020: 11–12). As a result, the number of border crossings into the EU, either regular or irregular, is expected to augment significantly. In parallel, new technologies, like avatars and virtual reality, which are currently being tested, will become new border control tools that could redefine borders and the border control mode. These portray a changing border control environment. It is not sure, though, whether they will also bring a new culture of border control in Europe. Regardless of any cultural evolution, namely either with Warsaw or with another culture of border control, Frontex has succeeded in consolidating its role in EU border control, acting confidently or, if you prefer, being self-assured about its position at Europe's borders. Is Frontex right in being self-assured?

Between a rock and a hard place?

With Frontex impacting European border control through culture, more questions arise about Frontex's function as well as its role in EU border control. On that account, far from solving all the enigmas on Frontex, the book might have created new ones.

A possible enigma concerns Frontex's governance. The first Executive Director, Ilkka Laitinen, led Frontex for 10 years, whereas its next Executive Director, Fabrice Leggeri, headed the agency for 7 years and 4 months. Both cases refer to appointments characterised by a long-term duration. On that account, one could wonder whether there is any correlation between Frontex's strategic planning fostered at the executive level and Frontex's cultural impact in EU border control. Put differently, would a shorter term of office of its Executive Director have lessened Frontex's cultural impact on the other members of the border control community or decayed Frontex's cultural traits? On qualitative characteristics, Laitinen, namely the first Frontex Executive Director, who set off Frontex, started his professional career serving as a border guard in the Finnish Border Guard. Leggeri, though not a border guard, was actively involved with border control and irregular migration during his work at the French Ministry of the Interior. Following that, one may wonder whether a different selection, that is, the appointment as Frontex's Executive Director of a person less familiar with border control matters would have led Frontex on a different path or would have triggered a shift in Frontex's cultural traits. The above questions unavoidably lead to a more general concern, namely who governs Frontex? Is it its Executive Director, the Executive Director with the three Deputies, the management board, the heads of Frontex's divisions and centres, or maybe it is the whole staff? Solving the enigma of Frontex's governance is not an easy task. Therefore, it requires a new research scrutiny solely focused on answering some of the above-mentioned preoccupations.

Another enigma refers to Frontex's broader cultural impact. The book discussed Frontex's role in EU border control. Yet, besides border control, does Frontex impact on another policy field or area? For instance, these last years, it has been possible to witness an enhanced cooperation between Frontex and various actors in defence, such as NATO and CSDP missions. Similarly, whilst performing its coastguard functions, Frontex works closely with other EU agencies, namely the EMSA and the European Fisheries Control Agency (EFCA). Is this cooperation being accompanied by any cultural transmission? Does Frontex promote its cultural traits to other policy fields, like it has successfully accomplished with Warsaw culture in border control? To provide an answer to that riddle requires delving into each of these synergies so as to trace Frontex's function outside of its border control remit. That means, a fresh research endeavour and orientation.

Part of the above enigma that could possibly be fuelled by the findings presented in this book refers to Frontex's impact outside of the EU geographic area. As already mentioned, Frontex has prioritised its relations with non-EU countries, signing agreements, deploying liaison officers, participating in evaluation missions, opening risk analysis offices as well as conducting training and out-of-area operations. In fact, recently, Frontex celebrated its 3-year operational presence in Albania, which, according to Frontex, is delivering outstanding results and has 'marked a new phase for border cooperation between the EU and its partners in Western Balkans' (Frontex, 2022). Drawing on that, it might be of relevance to explore whether, via its cooperation with third countries, Frontex has started extracting its border control assumptions and practices or the border control model of the Warsaw culture beyond the EU space. Key to providing an answer to that enigma is research on Frontex's relations with third countries, especially neighbouring states with which Frontex has built a holistic operational cooperation or countries weighting as important for the successful conduct of Frontex return operations, namely the most commonly encountered countries of origin of returnees.

Another set of enigmas touch on the democratic control and accountability of Frontex. While those issues have been covered by the literature, the insights of this book regarding the role and impact of Frontex suggest that further research on the limitations of existing accountability mechanisms and possible reforms of the legitimation regime of Frontex is necessary. How can the European Parliament exert any real control over Frontex without taking account of the cultural dimension? How easy is it for non-border control practitioners — such as politicians — to understand the real impact of Frontex decisions? Is the setting up of internal controls within the agency the answer? Or maybe the European Parliament and the Council should start relying on external expertise, that is, experts outside of the border control community?

All of these enigmas are instrumental in opening up Frontex's scrutiny to fresh research paths. At the same time, they invite us to continue looking a bit more carefully at this EU agency, as it still has mysteries surrounding its function. After all, the preceding pages showed that Frontex is not just an actor but a rather complex one. So, it defies easy or conventional answers.

Looking ahead

With Schengen's death and Warsaw's rise, Frontex seems to have assured, at least for the time being, its central position in the border control community and, as a result, at Europe's borders. But what is the outlook for the future? Will it keep its central position, affirming therefore a non-decreasing impact on EU border control? Will it continue defining the model of border control pursued at Europe's borders? Will it remain being considered an essential border control actor? Although it is bold to talk about the future or formulate predictions for things to come, let alone in today's turbulent and ever-changing era, still, the book's findings could provide a valid indication about Frontex's future trajectory and the prospective development of border control in Europe.

In arguing that border control has recently moved from Schengen to Warsaw culture, this could indicate that border control might continue to be shaped by the Warsaw paradigm, at least for the foreseeable future. After all, certain major challenges arising for Europe's borders have not yet led to the emergence of alternative cultural paradigms of border control, as shown in Chapter 7. So, it is not daring to assume that Warsaw culture will maintain its dominant position in EU border control at least for the near future.

Following that, Warsaw's elements may become even more pronounced at the EU external borders, reconstructing gradually the border, as has already occurred, after Warsaw's retention, with the border control mode. Accordingly, Warsaw's border control assumptions and practices could become diffused from border control to the border, shifting the border towards becoming not just a 'locus' of border control assumptions and practices, but a fully-fledged constructor of these characteristics. That would equate to a border redefinition. By this token, borders might become more securitised, technocratic, and re-territorialised, infusing surveillance and intelligence.

As long as Warsaw culture remains being the dominant culture for border control in Europe, Frontex, which constitutes Warsaw's main promoter and enabler, is expected to continue occupying a key place and role at the EU external borders. In fact, with each passing day that Warsaw culture remains uncontested, Frontex solidifies its central position as a border control actor. In that context, Frontex's continuous enhancement is far from puzzling. By contrast, with Warsaw culture, Frontex is expected to continue to augment institutionally and operationally, acquiring a wider remit, new powers, and additional tasks.

Warsaw's dominance may change perhaps if significant limits are placed on Frontex's function. Limits on Frontex may impede the nourishing by Frontex of the Warsaw culture as they would evoke a restricted role for the agency both within and outside the border control community losing therefore its freedom to pursue and promote Warsaw culture.

Frontex's dominance may be challenged if Warsaw culture becomes defeated by alternative narratives of border control, not supported by Frontex. Actually, this scenario is not that likely. At the risk of becoming side-lined in the event of a new culture of border control replacing Warsaw, Frontex most probably will adapt its strategic choices by trying to assert a role in the new culture. After all, Frontex has just one clearly defined goal: to ensure its institutional survival. Put differently, to continue existing. On that account, the promotion of Warsaw culture is a means to an end. The moment that Warsaw stops fulfilling Frontex's end of institutional survival, Frontex is expected to leave Warsaw's side, developing at the same time a new cultural trajectory or becoming a zestful supporter of any new regime with a strong potential for developing the new dominant border control culture.

Concluding reflections

Whilst writing the last pages of this book, Frontex is being thrust once again in the spotlight, after the decision of its Executive Director, Fabrice Leggeri, to resign well before concluding his second term of office, in the aftermath of the findings of an investigation conducted by OLAF regarding the agency's role in human rights abuses and instances of misconduct. What conclusion might be drawn, even implied, from its Executive Director's resignation is that Frontex has gained a fair share of authority over time, which leads it to get out of control. Nevertheless, the devotion to Frontex has not wavered, and most probably it will never do. Just a few days after Leggeri's resignation, the Commissioner for Home Affairs, Ylva Johansson, visited Frontex's headquarters to express her undeviating support towards the agency and its staff. Her starting words whilst addressing Frontex staff were: 'I feel proud of you. I am proud of your work', assuring them that 'Europe really needs a strong and well-functioning Frontex' (European Commission, 2022). So, despite the fierce criticism, Frontex continues doing its 'job', that is, to protect the border. But, as the book pointed out, it does more than that. Long ago, it transcended its technical role attributed by its creators. Promoting Warsaw culture, Frontex's chief function has moved from managerial tasks to a fundamental cultural impact shaping the EU borders and border control. Thereby, the need to re-evaluate Frontex's mission by taking into account all of its functions and enacting a system of internal and external checks and balances that until now seems to be missing from Frontex's function, so as to substantially improve its democratic accountability and oversight, is more relevant than ever. Otherwise, the agency will continue acting as it deems fit for its own needs and ambitions. That may lead to more intense agency involvement and therefore it could bring new challenges for both the EU borders and the EU project. After all, paraphrasing Marcus Aurelius, an agency's worth is no greater that its ambitions. So, rather than reckoning that Frontex will restrict itself to its technical tasks, we might start absorbing that it has not and most probably it will never do.

Note

1 Exception is a journal article that invokes Zaiotti's rationale to answer one of the research questions developed in the study (Frowd, 2014).

References

Agnew, J. (2008) 'Borders on the Mind: Re-Framing Border Thinking'. *Ethics & Global Politics*, 1(4), pp. 1–17.

Alvesson, M. (2013) *Understanding Organizational Culture*. London: SAGE.

Bellamy, R., Lacey, J. & Nicolaïdis, K. (2017) 'European Boundaries in Question?'. *Journal of European Integration*, 39(5), pp. 483–498.

Della Sala, V. (2017) 'Homeland Security: Territorial Myths and Ontological Security in the European Union'. *Journal of European Integration*, 39(5), pp. 545–558.

Diez, T., Stetter, S. & Albert, M. (2006) 'The European Union and Border Conflicts: The Transformative Power of Integration'. *International Organization*, 60(3), pp. 563–593.

European Commission (2022) 'Commissioner Johansson's Address to Frontex Staff'. Available at https://ec.europa.eu/commission/commissioners/2019-2024/johansson/announcements/commissioner-johanssons-address-frontex-staff_en (accessed June 2022).

Frontex (2020) *Strategic Risk Analysis 2020*. Warsaw: Risk Analysis Unit.

Frontex (2021) *Results of Research & Innovation Activities 2020*. Warsaw: Frontex.

Frontex (2022) 'Three Years of Operation in Albania'. Available at https://frontex.europa.eu/media-centre/news/news-release/three-years-of-operation-in-albania-Uot4JP (accessed June 2022).

Frowd, P. (2014) 'The Field of Border Control in Mauritania'. *Security Dialogue*, 45(3), pp. 226–241.

Gray, C.S. (1999) *Modern Strategy*. Oxford: Oxford University Press.

Kier, E. (1997) *Imagining War: French and British Military Doctrine between the Wars*. Princeton: Princeton University Press.

Lefebvre, H. (1991) *The Production of Space*. Oxford: Blackwell.

Meyer, C.O. (2006) *The Quest for a European Strategic Culture: Changing Norms on Security and Defence in the European Union*. Hampshire: Palgrave Macmillan.

Newman, D. (2003) 'On Borders and Powers: A Theoretical Framework'. *Journal of Borderlands Studies*, 18(1), pp. 13–25.

Newman, D. (2006) 'Borders and Bordering: Towards an Interdisciplinary Dialogue'. *European Journal of Social Theory*, 9(2), pp. 171–186.

Newman, D. (2011) 'Contemporary Research Agendas in Border Studies: An Overview'. In: D. Wastl-Water (ed.), *The Ashgate Research Companion to Border Studies*. Farnham: Ashgate, pp. 33–47.

Newman, D. & Paasi, A. (1998) 'Fences and Neighbours in the Postmodern World: Boundary Narratives in Political Geography'. *Progress in Human Geography*, 22(2), pp. 186–207.

Paasi, A. (2009) 'Bounded Spaces in a "Borderless World": Border Studies, Power and the Anatomy of Territory'. *Journal of Power*, 2(2), pp. 213–234.

Regulation (EC) No 2007/2004 of 26 October 2004 Establishing a European Agency for the Management of Operational Cooperation at the External Borders of the Member States of the European Union [2004, OJ L 349/1].

van Houtum, H., Kramsch, O. & Ziefhofer, W. (eds.) (2005) *B/ordering Space*. Aldershot: Ashgate.

Zaiotti, R. (2011) *Cultures of Border Control: Schengen & the Evolution of European Frontiers*. Chicago: University of Chicago Press.

Index

Note: Page numbers in *italics* indicate figures and in **bold** indicate tables on the corresponding pages.

9/11 4, 25

ABC (automated border control) 137
accountability 5, 124, 184, 187
actor: institutional 131, 177; -ness 36, 46–47, 90, 101, 109; non-state 5, 8; policy 14; security 47, 51; social 39, 44, 48–49, 121; technocratic 135
Afghanistan 1, 95, **96**, 156
Africa(n), North 38, 68, 70, 74–75, 77, 84n1, 84n3, 107
Albania(n) 52, 71, 118, 152, 184; Greek-Albanian 93–94, 176
anchoring, culture's *11*–12, 126–128
Arab Spring 71, **72**, 84n3, **96**
assumption(s): cultural comparison 122–124; definition of border control 10–11; Frontex's 50–55, **59**; in Evros 102–105, **106**; in Lampedusa 79–81, **83**; Warsaw's 116–117, **119**
asylum 37, 91, 100, 129, 162; procedure 158, 164; seekers 5, 85n4, 110n12, 166, 183
Austria 71, 156–157, 176
autonomy 25–28, 31, 39, 180; *see also* independence

barrier(s) 3–4, 12, 13, 40n1, 46, 102, 105, 122, **123**, 151; *see also* border(s)
Balkan(s) 71, 90, 151, 161; Western 96, 184; *see also* WB-RAN
BCP(s) (border crossing points) 93, 102, 162
Belarus 1, **96**, 110n11, 128, 165
Bigo, D. 36–37
bilateral agreement(s) 71, 124, 152–154

border(s): as barriers 3–4, 12–13, 122–**123** (*see also* barrier(s)); -line(s) 52, 102, 116–117, 135; and as lines 67, 80, 107, 164; as being made 4–5, 67, 89, 176–177; creation of EU external 4, 24; studies 8; *see also* internal, border
bordering 4–5, 68, 91, 150, 166; co- 118; practices 38
borderland(s) 8, 53, 60n5, 102
BORTEC study 132
Brexit 14, 15, 48, 118, 147, 149–155, 168–169
Brussels 28, 138; culture 12–13, 122–125
Bulgaria(n) 33, 90–91, 93, 100, 109n1, 109n3, 110n9

Campesi, G. 135
case study 15–16, 68, 84, 90, 109
casus belli 163, 169–170n7
CEPOL (European Union Agency for Law Enforcement Training) 33
Chillaud, M. 37
CIRAM (Common Integrated Risk Analysis Model) 51, 53, 135
code of conduct, Frontex's 36, 51, 57, 119, 127, 130, 131
Colimberti, F. 27–28
Commission 7, 13, 25–26, 32, 37, 47–48, 56, 122, **123**, 127, 132–133, 137–140, 149, 157, 179, 181; President 1, 67, 89, 169n6; Commissioner 36, 186; *see also* Home Affairs
Common Core Curriculum 33, 51, 106
community, definition of border control 10

contestation (cultural) 9, 148;
  Schengen's 126–128, 139; Warsaw's
  148–149, 167–168; Westphalia's 12–13
cooperation, multilateral: challenges
  154, 160, 165, 167–168; comparison
  124–125; in Evros 105–**106**; in
  Lampedusa 82, **83**; as practice in
  Frontex 55–56, **59**; Warsaw's 118–**119**,
  123; Warsaw's condition 136
Council 7, 24, 27, 122, **123**, 127, 129,
  138–139, 150, 157, 169n2, 169n6, 185;
  Thessaloniki European 25
Court: of Auditors 23, 47; of Human
  Rights 71; of Justice 7, 17n1, 150, 169n2
COVID-19 155–161, 169n5; outbreak 1,
  45, 95, 155, 158, 162, 169; pandemic
  72–73, **96**, 159–161, 168
crime 34, 50, 52, 58; agency 152;
  cross-border *35*, 71, 76, **78**, 97, 98,
  101, 128, 150, 159; networks 54;
  organised 4
crisis, migration 1, **72**, 95, **96**, 116, 151,
  157, 166
CSDP (Common Security and Defence
  Policy) **78**, 79, 85n6, 184
cultural trait(s) 9, 39, 67–68, 107–108,
  137, 182; Frontex's 50–**59**;
  in Lampedusa 78–**83**; in Evros
  102–**106**
culture, importance and concept 6,
  8–10, 38–40; *see also* culture(s) of
  border control
culture(s) of border control: as
  analytical framework 10–14;
  application 115–119; *see also* Brussels;
  culture; Schengen culture; Warsaw;
  Westphalia
Cyprus 91, **96**, 102, 109n3, 162

debriefing: and debriefers 54, **78**, **101**,
  103, 106, 116; experts 83, 150; in Evros
  99, 105; Frontex 33, 34, 54, 58, 120; in
  Lampedusa 80–82
decision-making 26, 37, 160, 180–181
defence 58, 97, 184; Ministry of 75, 154
disembarkation 71, 73–74, 76, 80–82, 95
division(s), of Frontex *29*–*30*, 32–34; *see
  also* Frontex
dominant: culture of border control
  framework 12–14; Schengen 13–14,
  125, 138; Warsaw 126–128, 139–140,
  147–149, 168; Westphalia 12–13, 125
Draghi, M. 73

EBCG (European Border and Coast
  Guard) Day 46, 49, 130
ECRE (European Council on Refugees
  and Exiles) 6
effectiveness 12, 23, 55
EIBM (European Integrated Border
  Management) 34, *35*, 120; as strategy
  54, 118, **123**, 128, 133–134, 137;
  see also IBM
emergency 33, 34, 80, 153, 163;
  discourse 80; health 155, 157–161;
  measures 116, 124; state **72**, 107;
  see also RABIT
EMSA (European Maritime Safety
  Agency) 56, 184
English Channel, as border 150–155;
  see also Brexit; France
enlargement 25
EPN (European Patrols Network) 46,
  76; *see also* Mediterranean, Coastal
  Patrols Network
ESTA (United States Electronic System
  for Travel Authorisation) 34
ethics 34, *35*, 38, 48, 51
ETIAS (European Travel Information
  and Authorisation System) 32, *35*,
  52, 116, 117, 118, 135, 140; Regulation
  125; Unit *30*, 33–34
EUAA (European Union Agency for
  Asylum) 56
EUBAM (EU Border Assistance
  Mission) 77, **78**, 85n6, 120
European Agenda on Migration 31, 125
EUROSUR (European Border
  Surveillance) system 34, *35*, 48,
  57, 83, 116, 135; application 53, 80,
  103; Fusion Services 55, 118, 132;
  Regulation **123**, 128, 131–132;
  see also surveillance
EURTF (EU Regional Task Force)
  76–**78**, 82, 99–**101**, 105
Evros: battle of 102, 162–165; border
  92–97; prefecture 91; river 90–93
Executive Director 26–27, *29*, *30*, 47,
  51, 54, 55, 100, 114, 130, 131, 133, 150,
  152, 183–184, 186; *see also* Laitinen;
  Leggeri
expertise 45, 48, 77, 103, 133, 181; external
  185; professional 27, 83, 106; technical
  7, 56, 134; technocratic 55, 116
External Borders Fund 48
External Borders Practitioners
  Common Unit 24–25

False and Authentic Documents Online (FADO) system 31
fence 72, 151; Evros border 95, **96**, 97–98, 103, 107, 164
Fernández-Rojo, D. 37
FLO (Frontex Liaison Office/r) *30*, *35*, 77, **78**, 99, **101**, 136
fortress(es) 4, 118, 167
FOSS (Frontex One-Stop-Shop) 48, 57
FRA (European Agency for Fundamental Rights) 17n1, 31
FRAN (Frontex Risk Analysis Network) 46; *see also* risk analysis
France 13, 24, 33, 70, 71, 156, 157, 163, 176; UK-border 147, 151–155, 169; *see also* English Channel
Frontex: creation 25, 55, 119, 126; expansion *28–32*; governance 25–28, 183–184; literature 5–6, 36–38; mandate 2, 5, 23, 27, 31, 33–34, 37, 39, 79, 119, 128, 159, 169; organisational structure *29–30*, 32–34
fundamental rights 5, *30*, 32, 34, *35*, 180; as Frontex assumption 50–51, **59**, 121; *see also* human rights; FRA

Germany 13, 24, 154, 156–157, 158, 160, 176
Gibraltar 48, 150; *see also* Brexit
globalisation 3–4, 12

habitus 78, 118, 130
harmonisation 33, 37, 40n1, 48, 127, 130, 137
Hermes (operation) 76, **78**, 80, 81
Home Affairs: Commissioner 36, 186; Committee 17n1; DG 7; policy area 24, 149; meeting 46, 129; *see also* Commission
Horii, S. 37
hotspot(s) 34, *35*, 40n4, 56, 71–73, 76–**78**, 100–**101**, 107, 109, 116
human rights 5–6, 26–27, 38, 50–51, 179, 186; *see also* fundamental rights

IBM (integrated border management) 119, 128, 133, 137, 141n1, 141n2; *see also* EIBM
identity 49, 122; control 33; European 13, 175
independence: institutional 25–26, 98; national 91, 165; *see also* autonomy

information gathering: COVID-19 160; in English Channel 154; in Evros 105, **106**; in Lampedusa 82, **83**; as practice in Frontex 57, **59**; Ukraine 167; Warsaw's 117–**119**, 123; Warsaw's condition 136
insiders 4, 24, 161, 176; *see also* outsider(s)
institutionalisation 37, 38; culture's 10–12, 128, 139–140, 147, 149, 177
instrument, Frontex as 6, 14, 37–38, 115, 120, 179, 182
integrity, territorial 3, 12, 122, 175; Greece's 91, 102, 108, 164; Italy's 79
intelligence: and AFIC (Africa-Frontex Intelligence Community) 77, **78**, 81; artificial 54, 58, 154, 160; as assumption in Frontex 53–54, **59**; community 77, 136; comparison 108, **123**–124; COVID-19 161; in English Channel 152–153; in Evros 99, 104–105, **106**, 164; in Lampedusa 81, **83**; officers 155; Ukraine 167; Warsaw's 117, **119**; Warsaw's condition 135–136, 140
internal, border: abolition 4, 13, 24; reintroduction 116, 156–159; UK-France 149–155
Interpol 47
IOM (International Organisation for Migration) 33

JORA (Joint Operations Reporting Application) 48, 57, 82, 103, 135, 136

Kaunert, C. 36–37
knowledge 27, 39, 46, 52, 55, 56, 58, 60, 90, 99, 117, 119, 181; border control 45; new 16, 36, 89; organisational 57; specialised 103, 116, 139

Laitinen, I. 23, 26, 183
Lampedusa: border *69*, 70–75; island 67–70, 74–75
law-enforcement 33, 47, 57, 58, 77, **78**, 97, 118
Leggeri, F. 26, 27, 114, 147, 175, 183, 186; *see also* Executive Director
legitimacy 37
Léonard, S. 36–37
Libya(n) 15, 68, 70–**72**, 74, 77–**78**, 80–82, 84n2, 84n3, 85n6, 107, 162

192   Index

Malta 69, 70, 84n1, 158
management board 25–27, 29–30, 55, 132–133; decision(s) 59, 77, 134, 150; meeting(s) 28, 40n2
Mare Nostrum (operation) **72**, 79, 83
Mediterranean 37, 38, 54, 70, 74; Central 71, 76, 77, 79, 84, 95; Coastal Patrols Network 132; countries 158; Eastern **72**, 94, 109n3, 162; Sea 67, 68, 80, 82
MEDSEA study 132
Memorandum(/-a) of Understanding 56, 82
militarisation: as assumption in Lampedusa 79–80, **83**; comparison 107–108, 168; in English Channel 153–154; in Evros 102, **106**, 163–164
mission(s), joint 36, 135, 158, 182; *see also* operation(s)
mobility: cross-border 4, 93, 95, 117, 126, 163, 175, 183; formal 70–71; irregular 24, 93–94, 97; of population 70; of workers 40n1; restriction of 155–159; transnational 3, 12
modus(/-i) operandi 71, 74, 94, 150, 152
Moldova 166

national authorities 6–7, 32, 35, 36, 45, 48, 50, 55–56, 115
National Coordination Centres 132, 136
NATO (North Atlantic Treaty Organisation) 32, 47, 56, 91, 165, 184
Nautilus (operation) 38, 76, **78**, 81
navy: Italian 75, 79; Libyan 77; UK Royal 154
Nobel peace prize 74

OLAF (European Anti-Fraud Office) 17n1, 27, 186
operation(s), joint 26, 28, 29–30, 33–35, 38, 45, 46, 49, 55, 76, 82, 98–102, 104, 105, 107, 118, 127–130, 136, 150, 166–167; *see also* missions
OSCE (Organisation for Security and Co-Operation in Europe) 46
outsider(s) 3, 4, 24, 59, 93, 140, 161, 175–176; *see also* insiders

Parliament, European 7, 17n1, 25–27, 47, 133, 138–139, 184–185
Paul, R. 37
peace 1, 24, 32, 147, 161
Perkins, C. 38
plan(s), operational 26, 34, 36, 46, 48, 80, 135

Poland 25, 32, 110n9, 130, 138, 156, 165, 166
policing: as practice in Frontex 58–**59**, 121; in Evros 105, **106**, 108
Poseidon (operation) 77, 98–99, **101**, 150
power, executive 31, 46
practice(s): best 35, 55, 57, 100, 137; cultural comparison **123**–125; definition of border control 10–11, 79; Frontex's 55–**59**; in Evros 105–**106**; in Lampedusa 82–**83**; Warsaw's 117–**119**
practitiocratic 120–**123**, 126–127, 129–131, 139, 140, 178
pre-frontier 52, 54, 55, 104, 132, 135, 139
process-tracing 15, 16, 115, 129, 140
professionalisation: in Evros 106; in Lampedusa 83; as practice in Frontex 56–57, **59**; Warsaw's 118–**119**, **123**, 136–137
pursuit, culture's 11–12, 39–40, 126–127, 138–139
push-back(s) 71, 84n2, 95, 164

Qaddafi, M. al- 70, 71

RABIT (Rapid Border Intervention Teams) 31, 34, 35, 51, 110n11, 165; in Evros 99, **101**, 103, 105, 107, 128, 163
RAND 54
Regulation (EC) No 2007/2004 25, 31, 139; *see also* Regulation, founding
Regulation (EU) No 1052/2013 **123**, 125, 128; *see also* EUROSUR, Regulation
Regulation (EU) No 2016/1624 **123**, 125, 128, 131, 132–133, 139, 153
Regulation (EU) No 2016/399 125; *see also* Schengen, Borders Code
Regulation (EU) No 2017/2226 125
Regulation (EU) No 2018/1240 125
Regulation (EU) No 2019/1896 **123**, 125
Regulation, founding 31, 150; *see also* Regulation (EC) No 2007/2004
Reid-Henry, S.M. 38
relocation scheme 71, 85n4, 166
responsibility 7, 13, 24, 31–33, 50, 52, 81, 89, 102, 110n12, **123**–124; shared 47, 50, 115, 119, 128, 132, 139
retention, culture's 11–12, 128, 139–140, 182, 185
re-territorialisation 37–38, 52, 59, 108, 178, 185

risk analysis **31**, *35*, 37, 53, 57, 81, 100, 107, 119, 127, 176, 181; cell(s) 54, 77; and document 52; model 133; network(s) 49; offices 184; publications 36, 39; and reports 53, 130; Unit *29*, *30*, 32, 51, 56, 152; *see also* CIRAM; FRAN; TU-RAN; WB-RAN
Rumford, C. 38
Russia(n) 56, 128, 156, 158; invasion 16, 148, 161, 165–166

Salvini, M. 73
SAR (search and rescue) *35*, 73, 75–**78**, 80–**83**, 107–108, 153
Scharpf, F. 37
Schengen: acquis 2, 25, 150, 169n2; Agreement 24, 40n2, 48, 134, 149; area 17n4, 103, 116, 147, 157; and -associated 99, 157; Borders Code 54, 118, 156, 157; countries 26, 118; culture 12–14, 120, 122–126, 138–140, 148, 178; evolution 126–128; Information System 149; initiative 136; institutionalisation 10; intra- 73, 96, 176; -land 3; regime 40, 119; territory 4; travelling to 32, 52
Schout, A. 37
SCIFA (Strategic Committee on Immigration, Frontiers, and Asylum) 24–25; SCIFA+ 24, 129
secondary movements *35*, 53, 100, 152; *see also* Schengen, intra-
second-line: checks 53; officers **101**, 150
securitisation: as assumption in Frontex 51–52, **59**; COVID-19 159–160; in English Channel 153; in Evros 102–103, **106**; in Lampedusa 80, **83**; literature 36–37; Warsaw's 116, **119**, **123**–124, 139; Warsaw's condition 135; *see also* security
security, hard 13, 148, 161, 165, 167, 169; barriers 151; border *35*, 75, **101**, 120, 141n1, 141n4; and concern(s) 52, 102, 124; context 179; and emphasis **123**; focus 80, 122; in- 24, 116; internal 2, 47, 53, 119; landscape 1; national 124, 156, 159, 175; non-traditional 32, 79; -oriented 139; problem 116, 153; research 83; threat(s) 4, 51–52, 103, 117; *see also* securitisation
selection, culture's *11*–12, 126–128
Single European Act 13, 40n1, **123**, 125

smuggling 5, 34, 54, 71, 73, 75, 95, 159, 166; networks 74, 104; *see also* trafficking
socialisation 9, 49, 121, 127, 130
sovereignty 3, 12, 80, 102, 115, 122, 124, 151, 154, 164
Spain 24, **72**, 150, 156
standing corps *30*, 31, 33, *35*, 36, 45, 77, 82, 99, 129, 179
Statement, EU-Turkey 95, **96**, 100, 110n4, 162
strategy(ies) 3, 26, 34, 38, 51, 104, 116, 148, 157, 158, 182; fundamental rights 50; national 133, 141n2, 160; *see also* EIBM
surveillance **31**, *35*, 45, 132, 185; activities 38, 52; as assumption in Frontex 54–55, **59**; challenges 152–153, 160, 165, 167; companies 58; in Evros 95, 97–99, **101**, 104, **106**; in Lampedusa 76–**78**, 80–81, **83**, 120; Warsaw's 117, **119**, **123**–124; Warsaw's condition 136; *see also* EUROSUR

tariff 74, 95, 152
technocracy: as assumption in Frontex 55, **59**; and border control 182, 185; challenges 160; in Evros 103, **106**; in Lampedusa 80, **83**; Warsaw's 116, 118–**119**, **123**–124, 140, 178; Warsaw's condition 135
technology: challenges 154, 160–161, 165; in Evros **106**; in Lampedusa 82–**83**; as IT 48, 55, 57–58, 103, 136; new 33, 107, 126, 183; as practice in Frontex 58–**59**; specialised 116; Warsaw's 118–**119**, **123**–124; Warsaw's condition 137; Warsaw's materialisation 127–128
terrorism 4, 12, 34, *35*, 57, 80, 81, **101**, 103, 157
Themis (operation) 76, **78**, 80–81
threat 32, 51–52, 53, 91, 116, 154, 155, 156, 164, 167, 182; cross-border 34; existential 124, 159; terrorist 103, 157
trafficking 71, 75, 94; *see also* smuggling
training: activities 37, **78**, **101**; common **31**, 34, *35*, 133; curricula 77, 83, 127; professional 57; provision of 45, 100; specialised 103, 106, 130; Unit *29*, *30*, 33
traveller(s) 52, 93, 117, 135, 156, 158, 176
Treaty: of Amsterdam 4, 24, 134, 138; of Lisbon 50; of Maastricht 13, **123**, 125; Protocol 149–150; of Rome 40n1; Sandhurst 152; of Westphalia 3, 12

Treaty establishing the European Community 25; *see also* Treaty, of Maastricht
Tunisia 15, 68, 70–**72**, 74, 77, **78**, 82, 84n3, 107
TU-RAN (Turkey-Frontex Risk Analysis Network) 100–**101**, 105, 136; *see also* risk analysis

Ukraine 1, 16, 33, 128, 147–148, 165–169
Unisys 133
us/them 8, 67, 166
USA (United States of America) / US 4, 34, 46, 56, 70, **96**, 156, 163, 164, 176

variation (cultural) *11*, 126, 148, 168
Vaughan-Williams, N. 37–38
visa 32, 52, 94, **96**, 156
vulnerability assessment *30*, 32, *35*

war 1, 3, 12, 14, 24, 25, 70, **72**, 91, **96**, 147–148, 151, 154, 160, 161–170

Warsaw 15, 25, 28, 44, 77; culture 14, 16, **119**–126, 138–140, 149, 153–155, 159–161, 164–165, 167, 177–179; evolution 126–128; formation 129–138; future 185–186
WB-RAN (Western Balkans Risk Analysis Network) 46, 117; *see also* Balkans; risk analysis
Westphalia: culture 12–13, 122–125, 138, 148, 177; meta- 135; regime 137; state 3, 134; *see also* Treaty, of Westphalia
White Paper 13
Wiermans, M. 37
Wolff, S. 37
Working Arrangement(s) 56, 77, 100, **101**, 150
Working Group 17n1, *30*, 134

Zaiotti, R. 10–13, 115, 122–**123**, 126, 128, 138, 148, 178, 187n1